RAND NATIONAL DEFENSE RESEARCH INSTITUTE

T0302527

Toward Improved Management of Officer Retention

A New Capability for Assessing Policy Options

Michael G. Mattock, Beth J. Asch, James Hosek, Christopher Whaley, Christina Panis

Prepared for the Office of the Secretary of Defense

For more information on this publication, visit www.rand.org/t/RR764

Library of Congress Control Number: 2014945625

ISBN: 978-0-8330-8665-5

Published by the RAND Corporation, Santa Monica, Calif.

Preface

Each branch of the U.S. armed services must actively manage officer retention to ensure meeting its officer personnel force requirements. Consequently, the U.S. Department of Defense (DoD) needs the capability to assess alternative policies to enhance the retention of officers. This capability should be founded on empirically based estimates of behavioral response to policy and recognize that members are forward-looking and take into account future opportunities and uncertainty, as well as the outcomes of past decisions and policies, when making current decisions. Further, the capability should enable DoD to simulate or predict the effects that alternative policies can have on officer retention, as well as the costs of those policies. Although such capabilities have been developed for enlisted personnel and for specific communities of officers, such as pilots, no capability exists to examine the retention and cost effects that alternative officer management policies can have on officers in each service. The Office of Compensation within the Office of the Under Secretary of Defense for Personnel and Readiness asked the RAND Corporation to develop such a capability.

This report documents our efforts to implement such a capability for officers and illustrates its use. We statistically estimate the parameters of a dynamic retention model of officer behavior, and we use the parameter estimates in a simulation model to help evaluate the effect that changes in compensation can have on the retention of officers, with a particular focus on those targeting midcareer officers, and to show how policies that change the retention behavior of these officers can also change the aggregate retention of the population of officers at earlier or later years of their careers. The model can also be used to gauge the effect of alternative policies to enhance retention, such as the Air Force's Aviator Continuation Pay program. In addition, we have created a spreadsheet version of the model that can provide quick estimates of the effect that bonuses, gate pays, and separation pays can have on retention in all years of service.[1] This report provides the mathematical foundations and the source code for the spreadsheet model, which should enable DoD analysts to replicate, apply, and extend the model as they see fit. (The spreadsheet model is also available on request from RAND's Forces and Resources Policy Center.)

This research was sponsored by the Office of the Under Secretary of Defense for Personnel and Readiness (P&R) and conducted within the Forces and Resources Policy Center of the RAND National Defense Research Institute, a federally funded research and development center sponsored by the Office of the Secretary of Defense, the Joint Staff, the Unified Com-

[1] Gate pay is pay that is given upon reaching a particular rank or year of service (or both). It does not imply any sort of obligation; it just is a pay contingent upon reaching a particular goal.

batant Commands, the Navy, the Marine Corps, the defense agencies, and the defense Intelligence Community.

For more information on the RAND Forces and Resources Policy Center, see http:/www.rand.org/nsrd/ndri/centers/frp.html or contact the director (contact information is provided on the web page).

Contents

Figures

Tables

Summary

The management of officer retention, particularly midcareer officers, is an ongoing concern for the U.S. armed services. Given that the U.S. military permits virtually no lateral entry, all midcareer and senior officers must enter at the bottom and be retained. In this report, we estimate the parameters of a behavioral model of officer retention decisions and use these parameters in a simulation model to help evaluate the effect that changes in compensation can have on the retention of officers. We also show how policies that change the retention behavior of midcareer officers can also change the aggregate retention of the population of officers at earlier or later years of their careers. The model allows us to simulate the effect of compensation associated with an increased service obligation, such as under the Army's Graduate School for Service program. We have created a spreadsheet version of the simulation model that can provide quick estimates of the effect of bonuses, gate pays, and separation pays on retention in all years of service (YOSs). We developed these capabilities in response to a request from the Office of Compensation for Personnel and Readiness. Similar capabilities have been developed for enlisted personnel and for specific officer communities, but no capability existed to analyze the retention and cost effects that alternative officer management policies can have on officers in each service.

Data, Model, and Estimates

In our behavioral model of officer retention, individual officers are assumed to compare present and future military compensation with present and future civilian compensation, taking into account both their constant underlying preference for a military career as compared with a civilian career, and present and future uncertainty regarding environmental disturbances that may affect their relative valuations of military and civilian life. Thus, we need data from both military and civilian sources to estimate the model.

Individual-level data on officers' initial service obligation, source of commission, and their retention decisions were obtained from the Defense Manpower Data Center (DMDC). These data were supplemented with information on military pay by YOS in 2009, obtained from the Directorate of Compensation within the Office of the Secretary of Defense (OSD). We measure military pay in terms of average Regular Military Compensation (RMC), a measure of military pay that is traditionally considered comparable to civilian earnings. RMC includes basic pay, allowances for subsistence and housing, and the tax advantage associated with getting these allowances tax-free. In the model, we deflate RMC appropriately for each officer cohort based on the change in RMC relative to civilian compensation (as reflected in the U.S. Bureau of Labor Statistics' Employment Cost Index [ECI]) over the time period covered by the data.

Our model includes the postservice civilian earnings of military personnel. Postservice earnings of veterans were derived from the Current Population Survey (CPS) by computing the 80th percentile of civilian pay for males aged 26–50 holding master's degrees, employed full time, full year, in management occupations. This group definition in terms of age, education, and occupational area corresponds to the general characteristics of military officers. We use the 80th percentile because tabulations indicate that officer RMC falls at about the 80th percentile of the earnings of civilians with these characteristics.

We also include military retirement benefits. We assume that all officers are covered by the system known as "high-three," introduced in 1981.[2] Under the military retirement system, members are vested at 20 years of service in an annuity that begins immediately and that is based on a formula that computes benefits based on annual basic pay and years of service. The high-three system bases the annuity on the individual's highest three years' average pay rather than the final year's pay. Military retirement benefits are computed based on information on years of service and annual basic pay. The latter was provided to us by the Directorate of the Office of Compensation in OSD.

The behavioral model is designed assuming that officers are forward-looking and take into account future opportunities and uncertainty when making current decisions. The model is also designed to enable the simulation of the effects that alternative compensation policies can have on officer retention; that is, the model is specified in terms of parameters in which estimates of those parameters do not depend on the particular compensation policy that was in effect at the time covered by the data. Instead, the parameter estimates reflect underlying characteristics of the distribution of preference for military careers on the part of the officer population, as well as the intrinsic uncertainties that officers face both in the military and in potential civilian careers.

As mentioned above, officers in the behavioral model make their decisions to stay or leave based on current and future compensation, their relative preferences for military versus civilian opportunities, and current and future environmental uncertainty. The basic model of officer behavior can be thought of as a simplified version of the model originally proposed by Gotz and McCall (1984). We estimate parameters using the basic model and later construct simulations that elaborate on the basic model by allowing officers to commit to an additional service obligation in exchange for bonus pay or an in-kind benefit.

Our parameter estimates for the Army conform to our expectations and previous results in the literature with respect to the relative taste for military service of those who entered the office corps through the Reserve Officers' Training Corps (ROTC) and the U.S. Military Academy at West Point. Where significant, the parameter estimates for the other services also

---

[2] Active-duty members are covered by one of three retirement systems that are each a defined-benefit system, depending on when they entered service. Those entering prior to 1981 are covered by a system in which the annuity formula bases retired pay on pay in the final year of service. Those entering between 1981 and 1986 are covered by the "high-three" system described in the text, and those entering after 1986 are covered under "REDUX," in which retired pay is based on "high-three" pay but those retiring at 20 years of service receive 40 percent of high-three pay, growing to 50 percent for those retiring at 30 years of service. In contrast, those under the pre-1981 plan and under "high-three" who retire with 20 years of service receive 50 percent of pay. The National Defense Authorization Act for Fiscal Year 2000 (Pub. L. 106-65, 1999) made a key change to military retirement. Those covered by REDUX are given the choice at YOS 15 to stay under REDUX when they become eligible for retirement and receive a $30,000 bonus, or revert back to the high-three system.

conform to our expectations and previous literature; however, the estimates for the Navy, Air Force, and Marine Corps give a value for the variance in taste that is not significantly different from zero. This is not a credible result because it implies that all officers from a particular source of commission are identical in their preference for military service. This problem is likely due to the data we use, which do not provide information on the initial service obligation or any follow-on obligations that the officers incurred. Because of this, we inferred an initial service obligation based on the source of commission; although this approach seems to have been adequate for the Army, it was less successful for the other services.

Exploring Policy Alternatives

We use the estimates produced for the Army and calibrated parameters for the Air Force to explore policy alternatives for enhancing retention of midcareer officers. The purpose of these policy explorations is to demonstrate the modeling capability and the ability to conduct "what-if" analyses of how officer retention would change under alternative compensation policies. The policies we consider focus on improving midcareer officer retention, though the policies could be broadened and the modeling capability is relevant to the entire officer force.

Our initial simulations focus on simple bonus and separation pay schemes; these pays are simple in the sense that they do not entail the officer making any commitments, e.g., to stay a certain length of time or to leave at a certain future year of service. We then turn to simulations of more-complex compensation schemes based on current programs in the Air Force and Army. All the simulations were conducted using spreadsheet versions of the behavioral model. Full source code for the spreadsheet versions of the model is provided in Appendix A.

Our simulations of the effects of simple bonus and separation pay schemes demonstrate two key features: (1) in this steady-state model, a bonus or a separation pay affects retention throughout an officer's career, and (2) a desired increase in retention can be achieved either through using a bonus, a separation pay, or a combination of bonus and a separation pay. In addition, the analysis shows that bonuses and separation pays can be combined so that there is no net effect on the number of officers reaching retirement eligibility. Retaining officers to midcareer does not necessarily imply that retirement liabilities would increase if officers were provided an incentive to separate earlier.

Our simulations also show that more-complex compensation schemes can also be used to retain officers to midcareer and beyond. Our first set of simulations shows the effect that changing features of the Air Force Aviator Continuation Pay (ACP) program could have on pilot retention. The ACP currently gives officers who have just completed their flight school training obligation the opportunity to commit to an additional five-year obligation in exchange for receipt of a $25,000-per-year bonus. Note that, because we could not reliably estimate the parameters for the Air Force, these simulations are based on parameters calibrated to recent experience with Air Force pilot ACP "take rates" and retention. Although calibration is useful for producing illustrative results, making statistical estimates of the parameters is preferable because they reflect behavior revealed in the data used to generate the estimates.

The simulations show the effect that varying either the bonus level or the ACP contract duration can have on both cumulative retention and ACP contract take rate. Increasing the bonus always increases the cumulative retention rate and take rate; however, increasing the contract duration does not always increase retention and take rate. We believe that this occurs because officers value flexibility; shorter contract durations are sometimes preferable to longer contract durations because they give more flexibility even though the cumulative bonus payment for shorter contracts is always smaller than that for longer contracts. These simulations are shown in Figures S.1 through S.4.

The Army provides an interesting example of a benefit associated with an agreement to incur a service obligation. The primary difference from the Air Force example above is that the Army service obligation is split into two parts: The first obligation provides an officer with the option, but not the requirement, to attend graduate school in exchange for a further service obligation of three years. It also contrasts with the Air Force example in that no benefit is paid during the additional three years of the initial service obligation, and a benefit is paid only if an officer decides to use (or exercise) the graduate school option after serving the full length of the extended initial service obligation.

Under the Army program, a newly commissioned officer can choose to extend his or her commissioning obligation by three years in exchange for the option to attend graduate school once he or she has completed the service obligation. If the officer exercises the option to attend graduate school, his or her obligation is extended to the time he or she takes to go to graduate school plus three times the time spent in graduate school. For example, if an officer attends two years of graduate school, then his or her total obligation is increased by eight years: two for school and six for a payback tour. This means that an officer who exercises the option and goes to graduate school for two years will be carried through to 14 to 16 years of service, depending on the length of his or her initial commissioning obligation.

Originally, there was no cap placed on the number of participants in the Army's Graduate School for Service program. As of 2010, the number of participants was capped at 300. Our final set of simulations shows what the steady-state effect of the Army's Graduate School for

Figure S.1
Simulated Effect That Changes in Annual Aviator Continuation Pay Bonus Would Have on the Cumulative Retention Rate for U.S. Air Force Academy Commissions

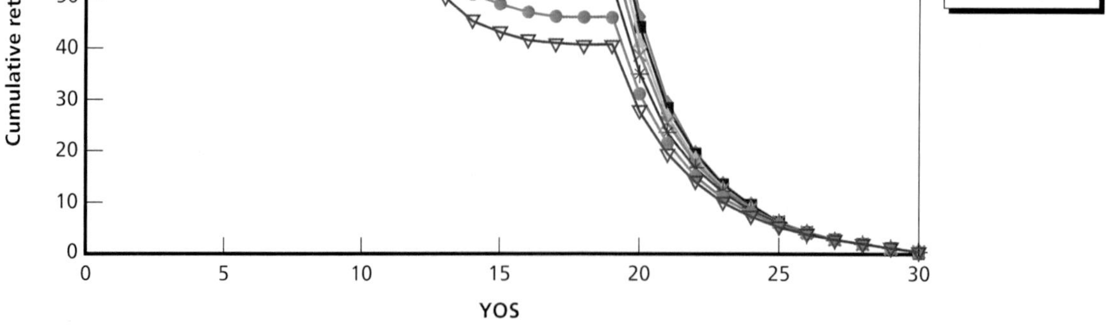

RAND *TR1260-S.1*

Figure S.2
Simulated Effect That Changes in Aviator Continuation Pay Contract Duration Would Have on the Cumulative Retention Rate for U.S. Air Force Academy Commissions

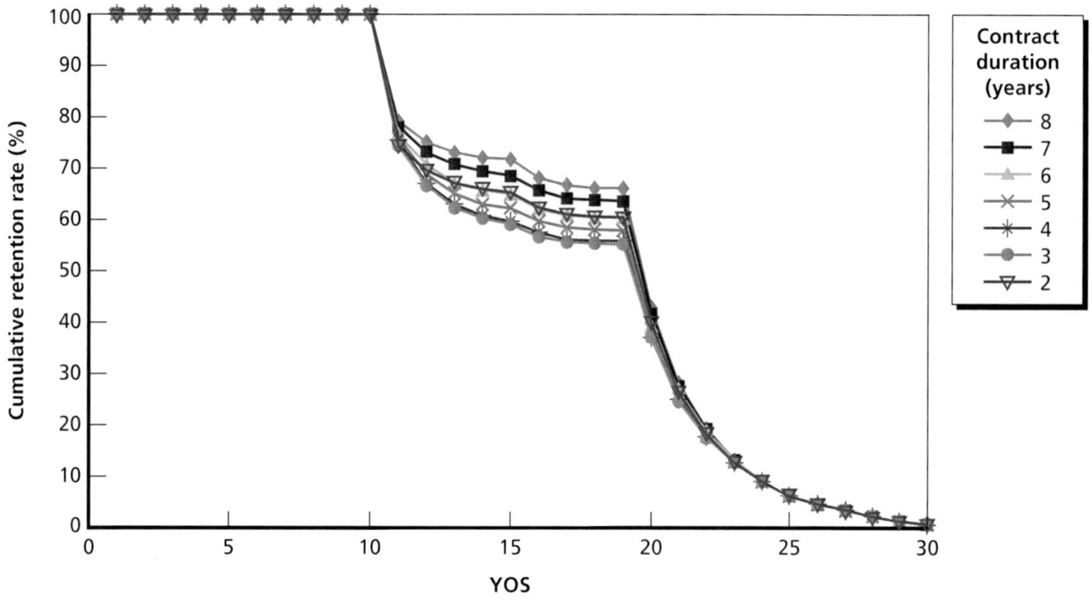

RAND *TR1260-S.2*

Figure S.3
Simulated Effect That Changing the Annual Aviator Continuation Pay Bonus Would Have on the Overall Take Rate

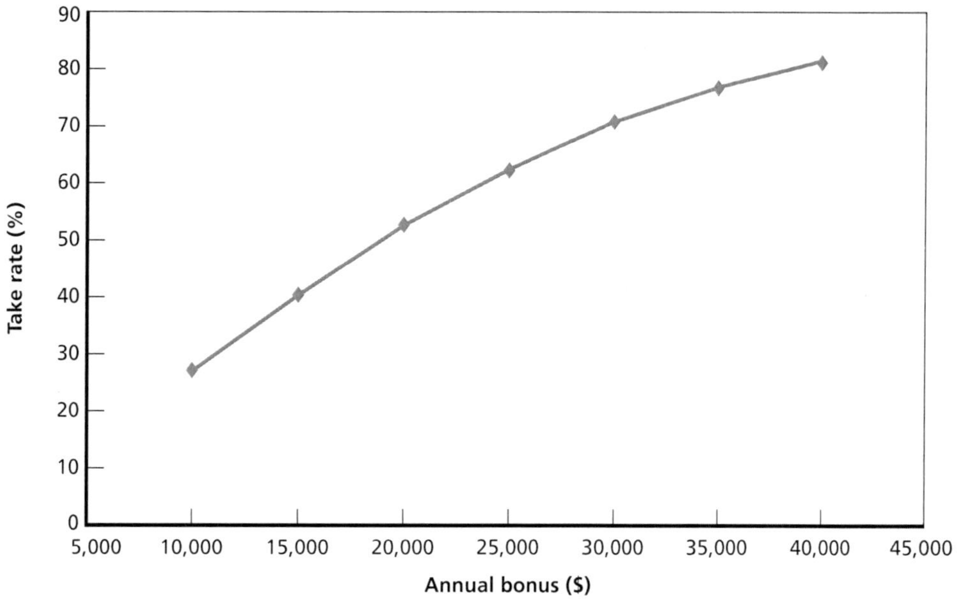

RAND *TR1260-S.3*

Service program would be if there were no cap on the number of participants, varying the present discounted value of the graduate school education and the increase in the lifetime stream of earnings. This is shown in Figure S.5. It is remarkable that even small values for graduate

Figure S.4
Simulated Effect That Changing the Aviator Continuation Pay Contract Duration Would Have on the Overall Take Rate

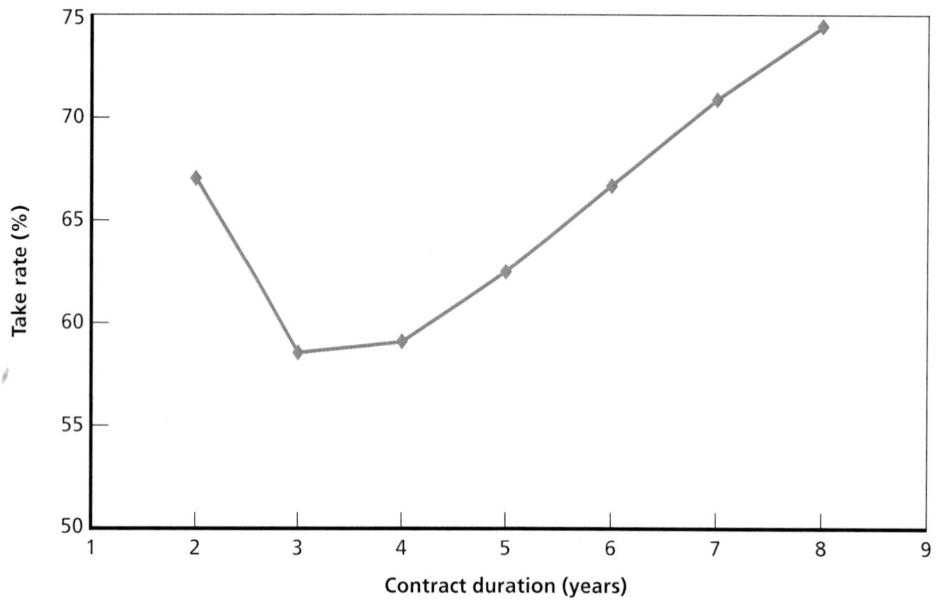

RAND *TR1260-S.4*

Figure S.5
Simulated Effect of the Army Graduate School for Service Program with No Cap on Number of Participants, Varying Present Value of Graduate School

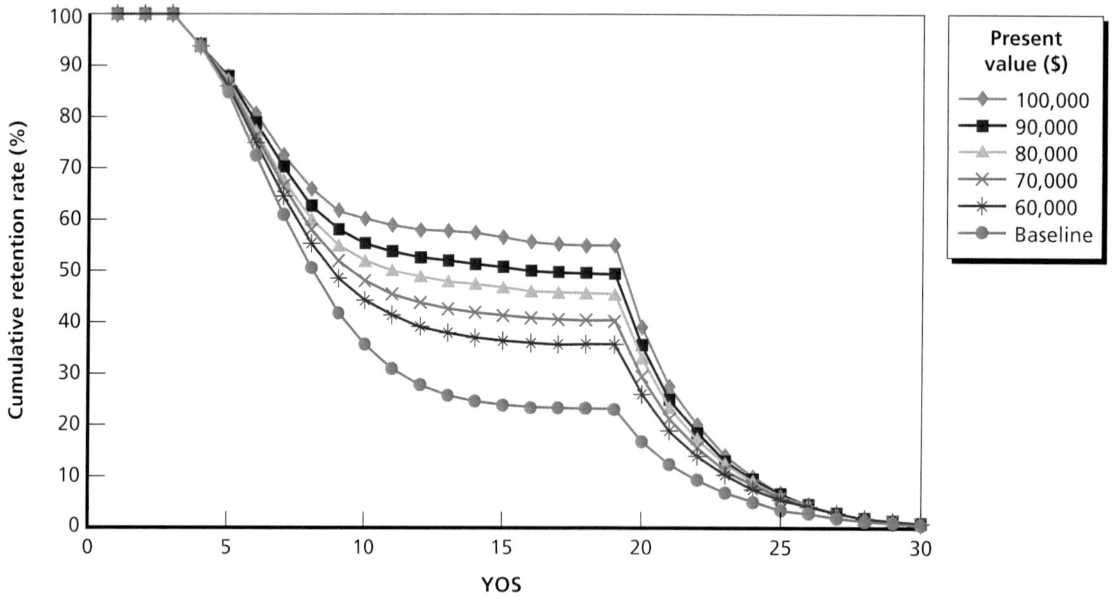

RAND *TR1260-S.5*

school produce relatively large retention results; however, it is important to bear in mind that these values do not reflect the full cost of this program to the Army. In addition, this example illustrates the prime difficulty with in-kind benefits of this type. Unlike bonus pays, which can be easily raised or lowered in response to officer behavior and the Army's needs, in-kind benefits cannot be easily changed. Thus, the only way to moderate the response to the Graduate School for Service program was to cap the number of participants.

Concluding Thoughts and Next Steps

To manage officer retention, policymakers require a capability that will enable them to conduct what-if exercises of different policy alternatives. Such a capability is needed by service, as well as for key officer communities. This report builds on past efforts for both officers and enlisted personnel to develop a dynamic model of officer retention behavior that provides such a capability. In the dynamic retention model, officer retention behavior depends on current and future opportunities in the military and in the civilian world, future uncertainty, and taste for military service relative to the civilian sector, with taste being allowed to differ across individuals. We estimate this model for the Army and then build a spreadsheet-based simulation capability that allows policymakers to conduct what-if analyses. The report summarizes the analyses and illustrates different policy options.

The policy simulations illustrate how the effect of a single bonus or separation pay reverberates throughout officer careers. Rather than rapidly diminishing over time, effects can propagate ten or 20 years in the future and can propagate to the beginning of an officer's career—that is, an officer at the beginning of his or her career will anticipate receiving a future bonus or separation pay.

The policy simulations show by example that the same retention target for a given year can be met either through a single bonus or separation pay or via a combination of a bonus and a separation pay. The simulations also show how multiyear commitments associated with either a cash or in-kind benefit can exert a powerful influence on officer decisions.

The results of the statistical estimation procedure show that the feasibility of this approach depends critically on the quality of the data, particularly data salient to the initial obligation faced by an officer and whether a stay-or-leave decision was voluntary or involuntary. We recommend that DMDC collect such data in the future from each of the services to facilitate statistical estimation of models of officer behavior. Meanwhile, calibration provides a useful, but not ideal, alternative approach.

The steady-state model we present can be used to gain useful insights into officer behavior over time. Merging this model of officer behavior with an equilibrium model of promotion or of officer inventory, or both, would be a useful direction to pursue in future work. In addition, further work to go beyond what-if—to come up with, say, the cost-minimizing bonus and separation pay schedule to reach a given YOS profile—could provide additional insights to military compensation analysts and policymakers.

Acknowledgments

We are deeply grateful for the support we have received from our sponsors throughout this project. With greatest pleasure, we acknowledge our debt of gratitude to Steven E. Galing, Deputy Director, Military Compensation, Office of the Under Secretary of Defense for Personnel and Readiness (OUSD[P&R]), and Robert M. Simmonds, Deputy Director for Programs, Officer and Enlisted Personnel Management (OEPM), OUSD(P&R). We are also indebted to the two reviewers of our draft report, Erik Meijer at RAND and John Warner, professor of economics (emeritus) at Clemson University.

Abbreviations

ACP	Aviator Continuation Pay
ADSO	active-duty service obligation
BAH	basic allowance for housing
BAS	basic allowance for subsistence
BRADSO	branch of choice for service
CPS	Current Population Survey
DMDC	Defense Manpower Data Center
DoD	U.S. Department of Defense
DRM	dynamic retention model
ECI	Employment Cost Index
FY	fiscal year
GRADSO	Graduate School for Service
OCS	Officer Candidate School
OSD	Office of the Secretary of Defense
OUSD(P&R)	Office of the Under Secretary of Defense for Personnel and Readiness
PADSO	post of choice for service
PDV	present discounted value
QRMC	Quadrennial Review of Military Compensation
RMC	regular military compensation
ROTC	Reserve Officers' Training Corps
USAFA	U.S. Air Force Academy
USMA	U.S. Military Academy
VBA	Visual Basic for Applications

WEX	Work Experience File
YOS	year of service

CHAPTER ONE

Introduction

Each branch of the U.S. armed services faces officer retention issues. For example, at present, the Army is focusing on the goal of increasing the retention of company-level officers in response to smaller officer accessions a decade ago, lower junior officer continuation rates in recent years, and force structure growth. The Marine Corps' force structure growth has also resulted in a shortfall of junior officers relative to required levels. Though the Navy has been downsizing in recent years, it is looking to increase its fleet size and shifting to a posture in which Navy personnel spend more time at sea. The Navy is placing increased emphasis on the role of midgrade officers in manning ships. Furthermore, both the Navy and the Air Force have experienced critical shortages among officers who are health care professionals and in special operations while reducing their force structure.

To address retention problems, the services often rely on different types of military compensation policy levers. Military compensation is a large fraction of all defense expenditures, which themselves account for the largest single discretionary spending item in the U.S. government budget. Compensation is the key tool for attracting, retaining, and eventually separating the right number and quality of people for the military.

Decisionmakers are often interested in conducting "what-if" analyses of the effects that different pay options can have on retention and the trade-off between retention and cost. For example, many policy alternatives could affect the retention decisions of midcareer officers: changes to retirement benefit levels and eligibility, changes to medical benefits, and new types of bonuses. Thus, decisionmakers in the U.S. Department of Defense (DoD) need a capability to understand, in an accurate and timely manner, the potential effects that policy alternatives can have on cost and retention. Although such capabilities have been developed for enlisted personnel and for specific communities of officers, such as pilots, no capability exists to examine the retention and cost effects that alternative officer management policies can have on officers in each service. The Office of Compensation within the Office of the Under Secretary of Defense for Personnel and Readiness (OUSD[P&R]) asked the RAND Corporation to develop such a capability. This report summarizes the findings of that study.

The capability to assess alternative compensation policies should be founded on empirically based estimates of behavioral response to policy and recognize that members are forward-looking and take into account future opportunities and uncertainty, as well as the outcomes of past decisions and policies, when making current decisions. Further, the capability should enable DoD to simulate or predict the effects that alternative policies can have on officer retention, as well as the costs of those policies. For the most part, past studies that have developed such capabilities have focused on the enlisted force. Studies that have focused on officers include the seminal work by Gotz and McCall (1984) and, more recently, a study of Air Force pilots (Mattock and Arkes, 2007). However, no recent study has focused on the entire officer force in each service.

This report documents our efforts to implement such a capability for officers. We statistically estimate the parameters of a model of officer behavior and use the parameter estimates in a simulation model to help evaluate the effect that changes in compensation can have on the retention of officers, focusing on policies targeted at officers in their midcareers. This analysis shows how policies that change the retention behavior of these officers can also change the aggregate retention of the population of officers at earlier or later years of their careers. The simulation model can also be used to gauge the effect of compensation or benefits associated with an increased service obligation, such as under the Army's Graduate School for Service program. This is a valuable feature of the model because incentives, such as those in the Army's program, are available only if the member agrees to an additional obligation of service, and the simulation model is capable of capturing the implicit cost to the member of "locking into" an additional obligation. In addition, we have created a spreadsheet version of the simulation model that can provide quick estimates of the effect that bonuses, gate pays, and separation pays can have on retention in all years of service (YOSs).

Although the policy excursions we consider focus on midcareer officers, the modeling capability is relevant to the entire officer corps, and we could consider other policies targeted to junior, or to senior, officers. Furthermore, the purpose of the policy analysis is to demonstrate a capability rather than to assess a specific policy.

The report is organized as follows. Chapter Two discusses the data, the behavioral model of officer retention decisions, and the estimates. Chapter Three uses the estimates from Chapter Two (and, where appropriate, parameters calibrated to recent data) to explore policy alternatives for changing midcareer officer retention. Chapter Four wraps up the report with a discussion of the conclusions and some thoughts on future research. The appendix documents the spreadsheet version of the model, including full source code for the behavioral model of officer retention decisions.

CHAPTER TWO

Data, Model, and Estimates

This chapter discusses the model we constructed to estimate the parameters underlying individual officer behavior. These parameter estimates are used in the simulation model we constructed, which is discussed in the next chapter. Readers interested mainly in understanding the capabilities of the simulation model we constructed may wish to advance directly to the following chapter.

We first discuss the data we used to estimate the behavioral model and then present the behavioral model of officer retention, followed by a discussion of its assumptions and limitations. Following this, we present the parameter estimates for each of the services and the model fit, and we compare the parameter estimates with previous literature on dynamic retention models (DRMs) of officer behavior.

Data

In the behavioral model of officer retention, individual officers compare present and future military compensation with present and future civilian compensation, taking into account both their constant underlying preference for a military career as compared with a civilian career, and present and future uncertainty regarding environmental disturbances that may affect their relative valuations of military and civilian life. Thus, we need data from both military and civilian sources to estimate the model.

The military data include, for individual officers, their initial service obligation and source of commission, their retention decisions, and their pays and retirement benefits. The civilian data include civilian pay data for observationally comparable civilians.

We constructed a database on individual officer careers, merging two Defense Manpower Data Center (DMDC) data sources: (1) the Proxy-PERSTEMPO data set for source of commission, DoD occupation, and demographic variables, and (2) the Work Experience File (WEX) for service, separation date (if separated), and date of commission. We had to infer each officer's commissioning active-duty service obligation (ADSO) using source of commission because neither data source included this field. For example, for the Army, we assumed that U.S. Military Academy (USMA) graduates have an ADSO of five years, scholarship Reserve Officers' Training Corps (ROTC) graduates have an ADSO of four years, and nonscholarship ROTC graduates have an ADSO of three years. These assumptions were based on input we received from our sponsor in the Office of the Secretary of Defense (OSD) and input on officer career management, such as Army Regulation 350-100 (Department of the Army, 2007). We track each officer's annual retention decision for those with an ADSO ending between 1990 and 2008 (the first retention decision is assumed to be at the end of the initial ADSO, and officers are tracked through December 2008).

We also constructed a variable for military pay over an officer's career. For military pay, we drew on the OSD Directorate of Compensation "Selected Military Compensation Tables" for Regular Military Compensation (RMC) average across pay grade by YOS for 2009. RMC is traditionally considered the earnings metric comparable to civilian earnings and includes basic pay, basic allowance for housing (BAH), basic allowance for subsistence (BAS), and an adjustment deriving from the allowances not being subject to federal income tax. RMC in general depends on YOSs, pay grade, and dependent status, but pay grade and dependent status are omitted from our model. This simplification means that we do not include probabilities of promotion, up-or-out rules, marriage, and divorce or separation. Pay grades, promotion probabilities, and up-or-out rules were included in our model for the tenth Quadrennial Review of Military Compensation (QRMC) (DoD, 2008a, 2008b) but have been dropped here because the compensation changes under consideration are not aimed at changing promotion speed or up-or-out rules, and the model estimates more quickly without these features.

In the model, we deflate RMC appropriately for each officer cohort based on the change in RMC relative to civilian compensation (as reflected in the U.S. Bureau of Labor Statistics' Employment Cost Index [ECI]) over the time period covered by the data. RMC changed significantly relative to civilian pay. In the late 1990s, military pay was increasing more slowly than the ECI, but, with the passage of the National Defense Authorization Act for Fiscal Year 2000 (Pub. L. 106-65, 1999), military pay was restored to competitive levels and, since then, has increased more rapidly than the ECI. Annual basic pay increases have been 0.5 percentage point higher than the ECI increase, and the housing allowance has increased (and has been refined to adjust for differences in housing cost by locality).

Civilian pay was computed from the 2009 Current Population Survey (CPS) by computing the 80th percentile of civilian pay for males aged 26–50 holding master's degrees, employed full time, full year, in management occupations, a group that has observed characteristics similar to those of the officer corps and so is designated as the relevant reference group. Figure 2.1 shows the 20th, 40th, 60th, and 80th percentiles of civilians in management occupations with this postcollege education as compared with RMC. We chose the 80th percentile because RMC tracks the 80th percentile for officers younger than age 40, as seen in the figure. Using the 2009 CPS to estimate civilian pay embeds the assumption that military personnel consider the civilian earnings of those of similar age, education, and occupational area as their relevant civilian earnings alternative. It also assumes that personnel forecast civilian earnings at some future age based on what individuals with similar observed characteristics earn currently. A limitation of this approach is that it assumes that the age structure of earnings does not change over time. However, because our model is a steady-state model, and the focus of our effort is not to model civilian earnings, we do not consider this limitation a major drawback.

Model

The behavioral model assumes that officers are forward-looking and take into account future opportunities and uncertainty when making current decisions. The model is also designed to enable the simulation of the effects that alternative compensation policies can have on officer retention; that is, the model is specified in terms of parameters that underlie the responsiveness

Figure 2.1
Regular Military Compensation for Officers Compared with Wage Percentiles for Civilians in Management with Master's Degrees, 2009

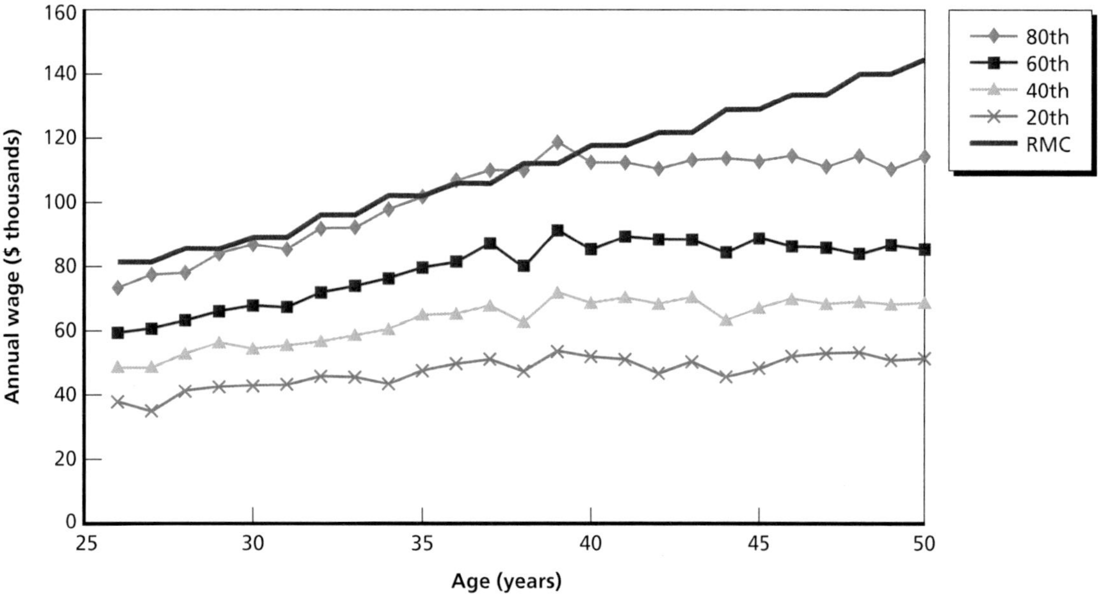

RAND *TR1260-2.1*

of personnel to compensation policy. In other words, rather than estimate parameters of the responsiveness of retention to compensation policies (e.g., elasticities), an approach that would result in estimates that embed the specific policies in effect over the course of the data period, we instead estimate parameters that reflect the distribution of preference for military careers on the part of the officer population, as well as the intrinsic uncertainties that officers face both in the military and in potential civilian careers. Of course, the distribution of officer preferences at entry toward the military could reflect compensation policy in effect at the time of entry, but our approach focuses on the underlying decision process of officers. As mentioned above, officers in the behavioral model make their decisions to stay or leave based on current and future compensation, their relative preferences for military versus civilian opportunities, and current and future environmental uncertainty.

The conceptual model of individual officer retention behavior is fundamentally simple. When an officer reaches a decision point, he or she compares the value of staying in the military with the value of leaving. If the value of staying is greater, then he or she stays. If the value of leaving is greater, he or she leaves. The officer faces multiple decision points in the course of a career, and, at each point, the officer makes an assessment of the value of staying in the military versus the value of leaving and chooses accordingly. Although this representation of how an officer might value either staying or leaving is simple, it can produce rich, interesting, and sometimes surprising behavior.[1]

We assume that each officer has some constant underlying value or "taste" for a year of military service. We also assume that each officer knows his or her taste, but we do not assume

---

[1] This model is based on the pioneering work of Gotz and McCall (1984). Additional studies include Asch and Warner (1994), Daula and Moffitt (1995), Hosek et al. (2004), and Asch et al. (2008). For a review of this work and related literature, please refer to Mattock and Arkes (2007).

that this taste is observed by the analyst. Instead, we assume that the initial population of officers will have tastes drawn from some distribution of tastes, as discussed more below. An individual officer will make decisions about staying or leaving conditioned on their taste. This means that officers who are otherwise identical in all observable factors might still behave differently because of their unique taste for military service. We use the symbol γ^M to denote this taste for military service. Table 2.1 describes all the symbols used.

During each year of military service, the officer also collects military wages and benefits. We denote the monetary equivalent of these wages and benefits by the symbol W_t^M. In this broad conceptualization, wages can include current monetary pay, such as basic pay, allowances, and special and incentive pays that military officers may receive, and benefits can include any type of military benefit, including retirement benefits, educational benefits, and health benefits. When we estimate the model, we use RMC for current pay because RMC makes up more than 90 percent of cash compensation (Asch, Hosek, and Martin, 2002), and we include retirement benefits but not health benefits or educational benefits.

Finally, during each year of military service, the officer is subject to a random environmental disturbance drawn from a stationary distribution. One can think of this disturbance as being the monetary equivalent of some a priori unknown event or collection of events that happen in a particular year—say, a good or bad assignment, or good or bad working condi-

Table 2.1
Explanation of Symbols in the Model

Symbol	Explanation
V_t^S	Value of staying for one more year
V_t^L	Value of leaving
β	Discount factor
γ	Net taste for military life, $\gamma^M - \gamma^C$
γ^M	Taste for military life
γ^C	Taste for civilian life
W^M	Military wage, inclusive of benefits
W^C	Civilian wage, inclusive of retirement benefits
R_t	PDV of the military retirement benefit payable to officers retiring in year t
ε_t^M	Military environmental disturbance at time t
ε_t^C	Civilian environmental disturbance at time t

tions beyond those normally expected in military life. We denote this environmental disturbance by the symbol ε_t^M.

Figure 2.2 gives a graphical depiction of the decision tree facing the officer. Both the individual taste and the environmental disturbance are drawn from continuous distributions, so they are represented by "fans" corresponding to the range of possible outcomes rather than by discrete branches. Once the taste is drawn, which occurs at the beginning and is known to the officer before the first retention decision is made, it stays with the officer throughout all future periods. The first fan shows the draw from the population taste distribution. Each officer's taste for the military is assumed to be known to the officer but unknown to the analyst. When the officer faces a stay-or-leave decision, he or she observes his or her draw from the environmental disturbance distribution (represented by the second and later fans) and can choose either to stay in the military and take the disturbance or to leave for civilian life and avoid the disturbance. The environmental disturbance distribution is assumed to be independent of the taste distribution. That is, the environmental disturbance in any period does not depend on whether the officer has a high or low taste.

In a given year of military service, the officer will collect utility, measured in dollars, that is the sum of these three elements: the constant underlying taste γ^M, the wages and benefits W_t^M, and the environmental disturbance ε_t^M, expressed as

$$\gamma^M + W_t^M + \varepsilon_t^M.$$

Similarly, on the civilian side, in a given year, the officer would collect utility, again measured in dollars, that is the sum of his or her constant underlying taste of civilian life, γ^C, the wages and benefits associated with a civilian career, W_t^C, and an environmental disturbance associated with civilian life in a particular year, ε_t^C:

$$\gamma^C + W_t^C + \varepsilon_t^C.$$

An officer does not consider only the current year when making decisions about staying or leaving. He or she also looks forward and considers the stream of utility he or she might receive as a military officer and compare that with the stream of utility he or she might receive

Figure 2.2
Decision Tree

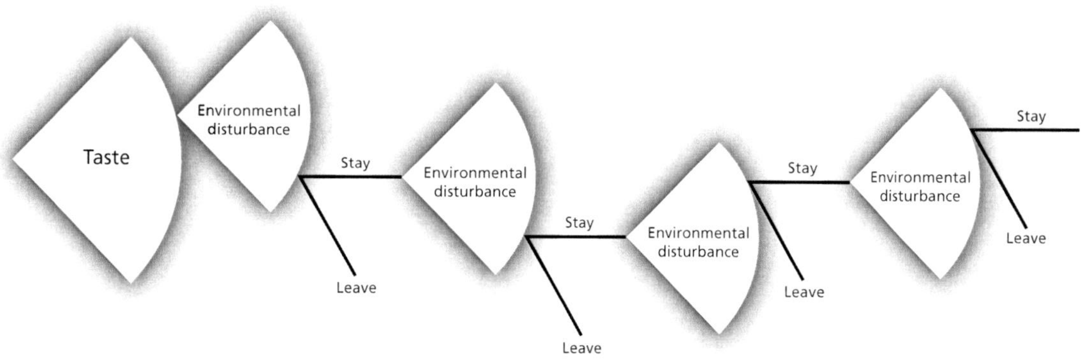

RAND *TR1260-2.2*

as a civilian. We use a discounted utility framework to value streams of utility, in which future utility is subject to a personal discount rate of r per year. For convenience, we also refer to

$$\beta = \frac{1}{1+r}$$

as a discount factor.

Thus, the discounted present expected value of the stream of utility from civilian life starting in year t is

$$\gamma^C + W_t^C + \varepsilon_t^C + \sum_{\tau=t+1}^{T} \beta^{\tau-1} E\left[\gamma^C + W_\tau^C + \varepsilon_\tau^C\right].$$

We can assume without loss of generality that the environmental disturbance terms have mean zero, so the above expression simplifies to

$$\sum_{\tau=t}^{T} \beta^{\tau-t}\left(\gamma^C + W_\tau^C\right) + \varepsilon_t^C.$$

The one missing element in computing the value of leaving is the value of the retirement benefit associated with leaving the military in year t. Let R_t denote the present discounted value (PDV) of the military retirement benefit if the officer retires in year t. Then we can write the value of leaving as

$$V_t^L = R_t + \sum_{\tau=1}^{T} \beta^{\tau-t}\left(\gamma^C + W_\tau^C\right) + \varepsilon_t^C.$$

The value of staying is slightly more complicated to write because we need to reflect the fact that the officer can revisit his or her decision to stay in the military in the following year. That is, in each period, the officer considers whether to stay or leave, whereas a civilian, having left the military, cannot reenter the military and always remains a civilian. If the officer stays in the military at year t, he or she will collect $\gamma^M + W_t^M + \varepsilon_t^M$ immediately and, in the following year, $t+1$ will be able to reoptimize based on the realizations of ε_{t+1}^M and ε_{t+1}^C, and the corresponding values of staying and leaving, V_{t+1}^S and V_{t+1}^L. We assume that the officer will choose the action that corresponds to achieving the maximum value, i.e., $\max\left(V_{t+1}^S, V_{t+1}^L\right)$. Taking the expected value and discounting by one year gives us

$$\beta E\left[\max\left(V_{t+1}^S, V_{t+1}^L\right)\right].$$

Thus, the value of staying is

$$V_t^S = \gamma^M + W_t^M + \beta E\left[\max\left(V_{t+1}^S, V_{t+1}^L\right)\right] + \varepsilon_t^M.$$

Notice that the value of military retirement enters the value of staying in period t through the expression for the expected value of the maximum, which contains the value of leaving in the next period.

We can further simplify the model, without loss of generality, by considering only the net difference in taste for civilian and military life, $\gamma = \gamma^M - \gamma^C$. With this simplification, the two equations describing the value of staying and the value of leaving become

$$V_t^S = \gamma + W_t^M + \beta E\left[\max\left(V_{t+1}^S, V_{t+1}^L\right)\right] + \varepsilon_t^M \tag{2.1}$$

and

$$V_t^L = R_t + \sum_{\tau=t}^{T} \beta^{\tau-1} W_\tau^C + \varepsilon_t^C. \tag{2.2}$$

Computing the Expected Value of the Maximum

If we assume that the environmental disturbance terms are bivariate normally distributed, then we can compute the expected value of the maximum as follows:

$$\overline{V}_t^S = V_t^S - \varepsilon_t^M$$

and

$$\overline{V}_t^L = V_t^L - \varepsilon_t^C.$$

If

$$\sigma \equiv \sqrt{\sigma_M^2 + \sigma_C^2 - 2\sigma_{MC}}$$

and

$$\upsilon \equiv \frac{\overline{V}_t^S - \overline{V}_t^L}{\sigma},$$

then

$$E\left[\max\left(V_t^S, V_t^L\right)\right] = \overline{V}_t^S \Phi(\upsilon) + \overline{V}_t^L \Phi(-\upsilon) + \sigma\phi(\upsilon),$$

where ϕ and Φ are the standard normal density and cumulative distribution, respectively (Clark, 1961).[2] Thus, for a given value of σ, we analytically can solve the recursive expression

[2] Asch et al. (2008) formulates the model differently by assuming that environment shocks have an extreme value distribution and taste has a normal distribution. The different shock distribution leads to a different form for the EMax expression.

forward for V_t^S, obtaining an expression for V_t^S in terms of the values of $\overline{V}_t^S$ and $\overline{V}_t^L$ in each future period. We assume that the disturbance terms are not autocorrelated. Furthermore, because $\overline{V}_t^S$ depends on $\overline{V}_{t+1}^S$, the model must be solved recursively.

We can interpret the final expression for $E\left[\max\left(V_t^S, V_t^L\right)\right]$ as the sum of the probability of staying, $\Phi(\upsilon)$, multiplied by the nonstochastic value of staying, $\overline{V}_t^S$, plus the probability of leaving, $\Phi(-\upsilon) = 1 - \Phi(\upsilon)$, multiplied by the nonstochastic value of leaving, $\overline{V}_t^L$, plus a term that gives the expected value of being able to choose the maximum of staying or leaving given the realization of the stochastic terms ε_t^M and ε_t^C. The value of this choice—or, stated differently, the value of having the option to choose—is greater the greater the variance of the environmental disturbance, i.e., the more uncertain is the future. As seen from the formula for σ, the sources of uncertainty are the variance in military outcomes, the variance in civilian outcomes, and the covariance between these outcomes. The value of σ is higher when military and civilian variances are higher and when military and civilian outcomes are negatively correlated (i.e., when the covariance term σ_{MC} is negative). Also, because the standard normal density has its highest value at zero, the value of this choice is also greater the closer the value of staying is to the value of leaving, i.e., the closer $\overline{V}_t^S - \overline{V}_t^L$ is to zero. At this point, the term $\phi(\upsilon)$ is at its maximum.

Deriving the Probability of Staying

The probability of staying in YOS t is equal to the probability that the value of staying, V_t^S, is greater than the value of leaving, V_t^L:

$$\Pr\left(V_t^S > V_t^L\right) = \Pr\left(\overline{V}_t^S + \varepsilon_t^M > \overline{V}_t^L + \varepsilon_t^C - \varepsilon_t^M\right) = \Pr\left(\overline{V}_t^S - \overline{V}_t^L > \varepsilon_t\right),$$

where $\varepsilon_t = \varepsilon_t^C - \varepsilon_t^M$ is normally distributed, because (1) both ε_t^M and ε_t^C are normally distributed with mean zero, (2) the probability density functions of normal distributions are symmetric about their mean, and (3) the sum of two normal distributions is also normally distributed. Then,

$$\Pr\left(\overline{V}_t^S - \overline{V}_t^L > \varepsilon_t\right) = \Phi\left(\frac{\overline{V}_t^S - \overline{V}_t^L}{\sigma}\right).$$

This implies that the cumulative probability of staying from YOS s to YOS t is

$$\prod_{\tau=S}^{t} \Pr\left(V_\tau^S > V_\tau^L\right) = \prod_{\tau=S}^{t} \Phi\left(\frac{\overline{V}_\tau^S - \overline{V}_\tau^L}{\sigma}\right).$$

This also implies that the cumulative probability of staying from YOS s to YOS $t-1$ and then leaving at YOS t is

$$\left(\prod_{\tau=S}^{t-1} \Pr\left(V_\tau^S > V_\tau^L\right)\right)\left(1 - \Pr\left(V_t^S > V_t^L\right)\right) = \left(\prod_{\tau=S}^{t-1} \Phi\left(\frac{\overline{V}_\tau^S - \overline{V}_\tau^L}{\sigma}\right)\right)\left(1 - \Phi\left(\frac{\overline{V}_t^S - \overline{V}_t^L}{\sigma}\right)\right).$$

Handling Unobserved Heterogeneity

We assume that the initial distribution of unobserved heterogeneity in the taste preference for the military, γ^M, is extreme-value distributed with mode α and scale δ. Other things equal, officers with higher taste for the military will have a greater propensity to stay than others; thus, the taste preference distribution will tend to change over time. Hence, we want to use only data for which we observe officers throughout their careers, and specifically including the beginning of the career, because our population-level distribution assumption holds for only the initial group of officers. We recognize that this group is itself affected by policy and economic conditions, but we have not modeled (and do not have data on) the recruitment of officers.

Further refinement of the model could allow the mode and scale parameters of the taste distribution to shift explicitly with demographic characteristics, such as gender or race, or with job characteristics, such as occupation, and we could estimate shifting parameters. Because of computational complexity, we do not include job and demographic characteristics in the model. Instead, we allow α and δ to shift with a key characteristics of officer service (namely, source of commission, and specifically whether an officer entered service through the ROTC program versus through a military academy).

To calculate the probability of observing an individual officer make a series of retention decisions, we will need to calculate the expected probability of staying given that the officer's taste was drawn from an extreme-value distribution, $f(\gamma)$, with mode α and scale δ.[3] We do this by integrating over the support of the probability distribution. By integrating over taste, we overcome the fact that we, as analysts, do not observe each officer's taste, and we obtain a probability expression that no longer depends on individual taste but rather on the parameters of the taste distribution, which we can estimate (Butler and Moffitt, 1982).

Given this assumption, the probability of staying from YOS s to YOS t is

$$\int_{-\infty}^{\infty}\left(\prod_{\tau=S}^{t}\Pr\left(V_\tau^S(\gamma)>V_\tau^L(\gamma)\right)\right)f(\gamma)d\gamma,$$

and probability of staying from YOS s to YOS $t-1$ and then leaving at YOS t is

$$\int_{-\infty}^{\infty}\left(\prod_{\tau=S}^{t-1}\Pr\left(V_\tau^S(\gamma)>V_\tau^L(\gamma)\right)\left(1-\Pr\left(V_t^S>V_t^L\right)\right)\right)f(\gamma)d\gamma.$$

3 The extreme-value distribution function is

$$\exp\left(-1\exp\left(\frac{\alpha-x}{\delta}\right)\right).$$

Its mean is $\mu = \alpha + 0.577\delta$, and its standard deviation is

$$\sqrt{\frac{\pi^2\delta^2}{6}} = 1.28\delta.$$

Constructing the Likelihood Function

We use maximum likelihood to estimate the parameters of the model. The likelihood function can be written as follows. We adopt the convention of writing $\Pr(Stay_\tau \mid \gamma_i, x_i, \theta)$ for $\Pr\left(V_\tau^S(\gamma_i, x_i, \theta) > V_\tau^L(\gamma_i, x_i, \theta)\right)$, where x_i is the vector of observables for officer i and θ is the vector of parameters we want to estimate: the standard deviation of the environmental disturbance, σ, the mode and scale parameters of the taste distribution, α and δ, and coefficients of any variables (e.g., an indicator variable for source of commission being ROTC) that are used to shift the mode and scale parameters.

For those observations that are censored, i.e., those observations for which the final state we observe is that officer i has decided to stay, we can write

$$L_i^{censored}(\gamma_i \mid x_i, \theta) = \int_{-\infty}^{\infty} \left(\prod_{\tau=S_i}^{t_i} \Pr(Stay_\tau \mid \gamma_i, x_i, \theta) \right) f(\gamma_i, \theta)\, d\gamma_i.$$

For those observations that are uncensored, i.e., those observations in which the final state we observe is that officer i has decided to leave, we can write

$$L_i^{uncensored}(\gamma_i \mid x_i, \theta) = \int_{-\infty}^{\infty} \left(\prod_{\tau=S_i}^{t_i-1} \Pr(Stay_\tau \mid \gamma_i, x_i, \theta)\left(1 - \Pr\left(Stay_{t_i} \mid \gamma_i, x_i, \theta\right)\right) \right) f(\gamma_i, \theta)\, d\gamma_i.$$

If we order our n observations so that the first k are the censored observations, then the likelihood function is

$$L(\theta) = \prod_{i=1}^{k} L_i^{censored}(\gamma_i \mid x_i, \theta) \prod_{i=k+1}^{n} L_i^{uncensored}(\gamma_i \mid x_i, \theta).$$

We can then find the parameter vector θ that maximizes the likelihood of observing the data in our sample. Again, we estimate the model via maximum likelihood.

Model Assumptions and Limitations

The model assumes that individuals are risk-neutral, rational, and forward-looking; differ in their taste for military service; and face common distributions of military and civilian shocks. Taste is assumed to be constant through time, and shocks are uncorrelated over time. The model does not treat promotion and does not account for deployment or for health benefits. The model uses an average civilian wage by age rather than a person-specific civilian wage.

Risk neutrality means that the model is not well suited to analyzing changes in the variance of military or civilian pay. This is potentially a limitation because civilian earnings variance has increased over time and the variance of military pay could change depending on military compensation policy. DeBacker et al. (2012) estimate cross-sectional earnings variance after controlling for age and education. The authors find little change in male earnings variance from 1989 to 1999, with variance in the range of 0.63 to 0.66, but, from 1999 to 2004, the final year of their data, earnings variance increased to 0.69. Military pay is highly stable, but its variance might have increased during the past decade because of the heightened pace of deployments and the increased use of bonuses. The core elements of military pay, namely, basic pay, basic allowance for subsistence, and basic allowance for housing, are paid according to published schedules and as mentioned account for over 90 percent of military pay on average (Asch, Hosek, and Martin, 2002). Military pay is higher during deployment, and, in general, the frequency, length, and conditions of deployment will affect the variance of military pay. Also, the increased use of bonuses during the past decade probably increased the variance of military pay. Still, in a military context, the payment of deployment pays and bonuses should not necessarily be viewed as an exogenous increase in pay variance. This is because these pays may be policy responses intended to stabilize retention when conditions of military service become more demanding. Thus, the pays may help to stabilize expected utility. By this reasoning, an increase in the nominal variance of military pay may mask underlying stability in the utility. By comparison, bonuses are also paid to keep military pay competitive with external opportunities (as opposed to a change in conditions within the military). In this case, higher external pay variance may drive higher military pay variance. The model does not account for the effects of higher pay variance on expected utility.

Models of decisionmaking typically assume rational behavior. However, the behaviorist school has drawn attention to other factors that affect behavior, e.g., the way an option to participate in a retirement savings program is framed (opt out versus opt in). Retention decisions might depend on "behavioral" factors, in which case the assumption of purely rational decisionmaking is too strong.

The assumption that individuals are forward-looking implies that individuals form expectations about future outcomes and are willing to act on them. In our case, the stability and predictability of military pay and careers support the formation of accurate expectations, and observed behavior in the data, namely, the increase in retention in years prior to the retirement vesting point at 20 years, is consistent with forward-looking behavior.

Differences in taste for the military can explain how two people, confronted with same expected value of their military career and civilian alternative and the same military and civilian shocks, might differ in their stay-or-leave choice. Individuals with higher taste for the military are more likely to stay in the force, and the model accounts for this selectivity. If instead the model assumed that all tastes were the same and there were no selection on taste over the military career, the estimate of mean taste would be biased upward and the model would tend to overpredict retention in the early career and underpredict it in the late career. Still, the model does assume that taste is constant over time, so the model does not treat the possibility that taste is affected by military experience. The model as currently specified fits the retention profile well, so the payoff to allowing individual taste to change over time might be small.

The assumption that military and civilian shocks are commonly distributed may be too strong. However, the empirical analysis allows the parameter estimates to differ by branch of service, and the model can be estimated for different groupings, e.g., by occupational area within a service, by occupational area across the services, and by demographic group. Thus, the assumption of common shock distributions can be explored empirically. Nevertheless, if the estimated shock variance is too high or too low for a group within an estimation sample, the group's EMax values will be too high; EMax increases with shock variance. This might lead to an overprediction of retention.

Deployments are known to affect retention (Hosek and Martorell, 2009; Fricker, 2002). Including deployment in the model could add to its explanatory power, but excluding it, as we do here, may have little effect on the parameter estimates. This is because deployment is statistically exogenous to active-duty personnel decisions (Lyle, 2006; Savych, 2008) and has little effect on military career progression as judged by the speed of promotion (Hosek and Totten, 2002).

Military pay enters the model as average pay by year of service, where the average is taken across personnel at different pay grades given the year of service. An earlier version of the model included promotion timing (the probability of promotion to a given grade at a given year of service) and up-or-out rules, but here the options we are considering are not aimed at changing promotion, so we omit it because the model estimates more quickly. A potential limitation is that, if a compensation or personnel policy change led to a large change in retention, then promotion speed would probably slow and, as a result, average military pay at a given year of service would decrease. Our use of average military pay assumes that it is unaffected by policy changes. If average pay fell, retention in preceding years would be expected to decrease.

The scope of the military health care benefit for active-duty personnel is broad and has remained so over time. However, the availability and scope of health benefit coverage offered by civilian employers has decreased over time, and the cost of health benefits has increased. These changes suggest that our use of average civilian wages overestimates the value of the civilian pay package in the later years of our data, e.g., after 2005, relative to the early years, e.g., 1990–1995. A civilian wage profile that factored in the value of health benefits offered by employers would probably increase with age at a slower rate than does the wage profile we use. This might cause an upward bias in the estimate for mean taste for military service. A refinement of the civilian wage profile to adjust for the changing value of the civilian health benefit can be pursued in future work.

The civilian wage used in estimating the model is the average civilian wage for full-time male workers, by age group. This is potentially a limitation because wages differ across individuals in a persistent way. Thus, having a civilian wage that incorporated more information about the individual, e.g., the individual's years of education, would be more accurate. But it is worth adding that our model could be first written to include an individual taste and an individual wage component, both constant. Because the taste and the wage components are both constant, this model is the same as a model with just a single taste variable and no idiosyncratic wage component, which is what we have. The implication is that our taste specification can be thought of as including a constant absolute civilian wage component for the individual—a component that is not identifiable. More-refined civilian wage data would capture part of the idiosyncratic component of wage, but there still would be a remaining idiosyncratic component, the reason being that an imputed wage would contain only part of what is idiosyncratic. This argument can be extended to include a military pay component, e.g., for a service member who is routinely promoted more quickly, or more slowly, than average and thereby earns more, or less, than average military pay given years of service. On net, our specification embeds a permanent wage component in the taste variable; it does not identify the wage component, but, to some degree, it does control for (capture the influence of) the wage component.

Before moving on to discuss estimates of the model parameters for each service, we want to alert the reader that Chapter Three illustrates the capability of extending the estimated model to simulate the effects of a variety of different policies. The simulations in general require adapting the value functions to reflect the policy being simulated, and the specific cases in Chapter Three include a retention bonus, a separation pay, an across-the-board pay increase, a compensation offer tied to a multiyear service commitment chosen by the individual (Air

Force Aviator Continuation Pay [ACP]), and an option to attend graduate school in exchange for accepting an obligation of three additional years of service (Army Graduate School for Service [GRADSO] program).

Estimates

The estimated coefficients for each service are shown in Table 2.2. The occupations used to estimate each of the parameters for each service are noted along with the estimated parameter values. In general, we exclude occupational areas that are eligible for large special and incentive pays, such as those in the medical fields and pilots, because our measure of military compensation does not incorporate these pays. These data sets included only officers whose source of commission is either a service academy or ROTC because Officer Candidate School (OCS) officer data seemed to include officers with prior service whose eligibility for retirement in a particular year could not be determined.

The estimates for the Army show that the mode of the taste distribution (α) is negative, as expected based on past research findings, but higher for officers whose source of commission is ROTC. Using the formula for the mean of the extreme-value distribution $(\mu = \alpha + 0.577\delta)$, we find mean taste of –34.53 for the Army, –10.90 for the Navy, –8.95 for the Air Force, and –10.55 for the Marine Corps. The standard deviation of the taste distribution is 1.28 times the scale parameter, δ. We find that the taste standard deviation is lower for those whose source of

Table 2.2
Coefficient Estimates

Coefficient	Army	Navy	Air Force	Marine Corps
α, mode of taste distribution	–50.63*	–12.95*	–10.83*	–11.16
	(23.44)	(0.87)	(2.70)	(81.71)
δ, scale parameter of taste distribution	27.90*	3.55*	3.25	1.05
	(10.02)	(1.27)	(13.66)	(225.31)
σ, standard deviation of environmental disturbance term	161.39*	37.99*	122.45	175.14
	(90.07)	(10.41)	(76.13)	(442.75)
$\alpha \times$ ROTC	16.94*	4.41	1.86	–3.45
	(8.59)	(71.32)	(338.21)	(114.94)
$\delta \times$ ROTC	–4.92	–3.58	–3.31	8.81
	(3.73)	(130.19)	(636.48)	(210.18)
β, discount factor	0.8875*	0.8255*	0.8857*	0.8845*
	(0.0436)	(0.0272)	(0.0549)	(0.2508)
n	11,754	2,250	2,591	917
log likelihood	–20,635	–4,524	–4,823	–1,838
DoD occupation codes	3xx–4xx, 7xx–8xx	3xx–4xx, 7xx–8xx	3xx–4xx, 7xx–8xx	3xx–4xx, 7xx–8xx

NOTE: * = coefficient that is statistically significant at 0.05. Standard errors are in parentheses.

commission is ROTC. The negative values of modal (or mean) taste are offset by the relatively high level of officer compensation, which also helps to buffer against the environmental disturbances. The estimates for the Army show a standard deviation of the environmental shock of $161,390, which is much larger than the values of the mode and scale of the taste preference distribution. This points to the large role played by environmental shocks. In some sense, this is a reflection of the extremely limited nature of the data describing the officer's career and his or her civilian job opportunities. It also means that, even with military pay set at a relatively high level, some officers will experience a large negative environmental disturbance and choose to leave. These findings for the Army conform with the results in Gotz and McCall (1984).

In contrast, the coefficient estimates for the scale parameter of the taste distribution for the Air Force and Marine Corps are not statistically significant. This may, in part, be due to the smaller sample sizes; Mattock and Arkes (2007) were able to generate statistically significant parameter estimates for rated Air Force officers with fewer than 2,000 observations. A more likely reason is that the Air Force and Marine Corps data from DMDC do not give a full accounting of the service obligations under which officers may find themselves in addition to that for service academy or ROTC. The data used in Mattock and Arkes were provided by the Air Force and included detailed service obligation information (e.g., the obligation incurred by attending flight school). In addition, some of the data cover the drawdown period in the 1990s, and the recent "force shaping" period in the Air Force (fiscal year [FY] 2006 to FY 2007). Further, in work currently under way using an active/reserve DRM, we find that including reserve participation in the structure of the model and data leads to estimates of active-duty taste variance that are statistically significant for all services. The discount factor is also estimable. This suggests that the additional empirical variation coming from reserve participation after active-duty service facilitates the estimation of the model parameters.

We have also estimated the model assuming that there is no unobserved individual heterogeneity. These coefficient estimates are reported in Table 2.3. In this case, the mode equals the mean, and we find that the estimated means are close to those from the model allowing unobserved individual heterogeneity given in the text above. However, the log likelihoods of the Army and Navy models are lower than for the corresponding models assuming unobserved individual heterogeneity, indicating that accounting for unobserved heterogeneity provides a superior fit to the data.[4] The coefficient corresponding to the observed heterogeneity (that is, the source of commission being ROTC) is significant in the Army and the Navy models and shows the expected effect, namely, that the population of officers whose source of commission is ROTC has a higher mean taste than the population of officers whose source of commission is a service academy.

Figure 2.3 shows graphs of the model fit for the coefficient estimates given in Table 2.2. The small circles are actual data of cumulative office retention to each year of service, and the lines are predicted retention. As seen, the model predictions fit the data well. An interesting aspect of the fit graphs is the predicted decrease in cumulative retention at YOS 20. This prediction is in a range beyond our data, which follows officers to YOS 18. However, the decrease is predicted by the model and is a consequence of the retirement benefit system. Retirement vesting occurs upon completing 20 years of service, and, once vested, a service member may leave the military and begin drawing retirement benefits. The availability of retirement benefits

---

[4] A likelihood ratio test shows that the fit is significantly better for the Army at the 1-percent level and for the Navy at the 10-percent level.

Table 2.3
Coefficient Estimates Assuming No Unobserved Individual Heterogeneity

Coefficient	Army	Navy	Air Force	Marine Corps
α, mode of taste distribution[a]	−37.28*	−10.52*	−10.91*	−14.10*
	(4.58)	(0.45)	(3.34)	(7.27)
σ, standard deviation of environmental disturbance term	167.26*	40.29*	150.78*	173.52*
	(24.06)	(5.00)	(48.63)	(88.67)
$\alpha \times$ ROTC	6.59*	1.68*	0.00	3.44
	(0.91)	(0.40)	(0.97)	(2.44)
β, discount factor	0.9341*	0.8328*	0.8988*	0.8955*
	(0.0060)	(0.0075)	(0.0176)	(0.0294)
n	11,754	2,250	2,591	917
log likelihood	−20,714	−4,526	−4,823	−1,838
DoD occupation codes	3xx–4xx, 7xx–8xx	3xx–4xx, 7xx–8xx	3xx–4xx, 7xx–8xx	3xx–4xx, 7xx–8xx

[a] The mode is also equal to the mean.

NOTE: * = coefficient that is statistically significant at 0.05. Standard errors are in parentheses.

as of YOS 20, but not before, increases the opportunity cost of military service and accounts for the decrease in retention at that point. In research currently under way, with data extending to YOS 21, the model's prediction is in accord with the actual data.

These estimates can be used to simulate the effects on retention of alternative policies. In Chapter Three, we discuss how we do policy simulations and present some examples.

Figure 2.3
Fit Graphs for Model with Individual Unobserved Heterogeneity

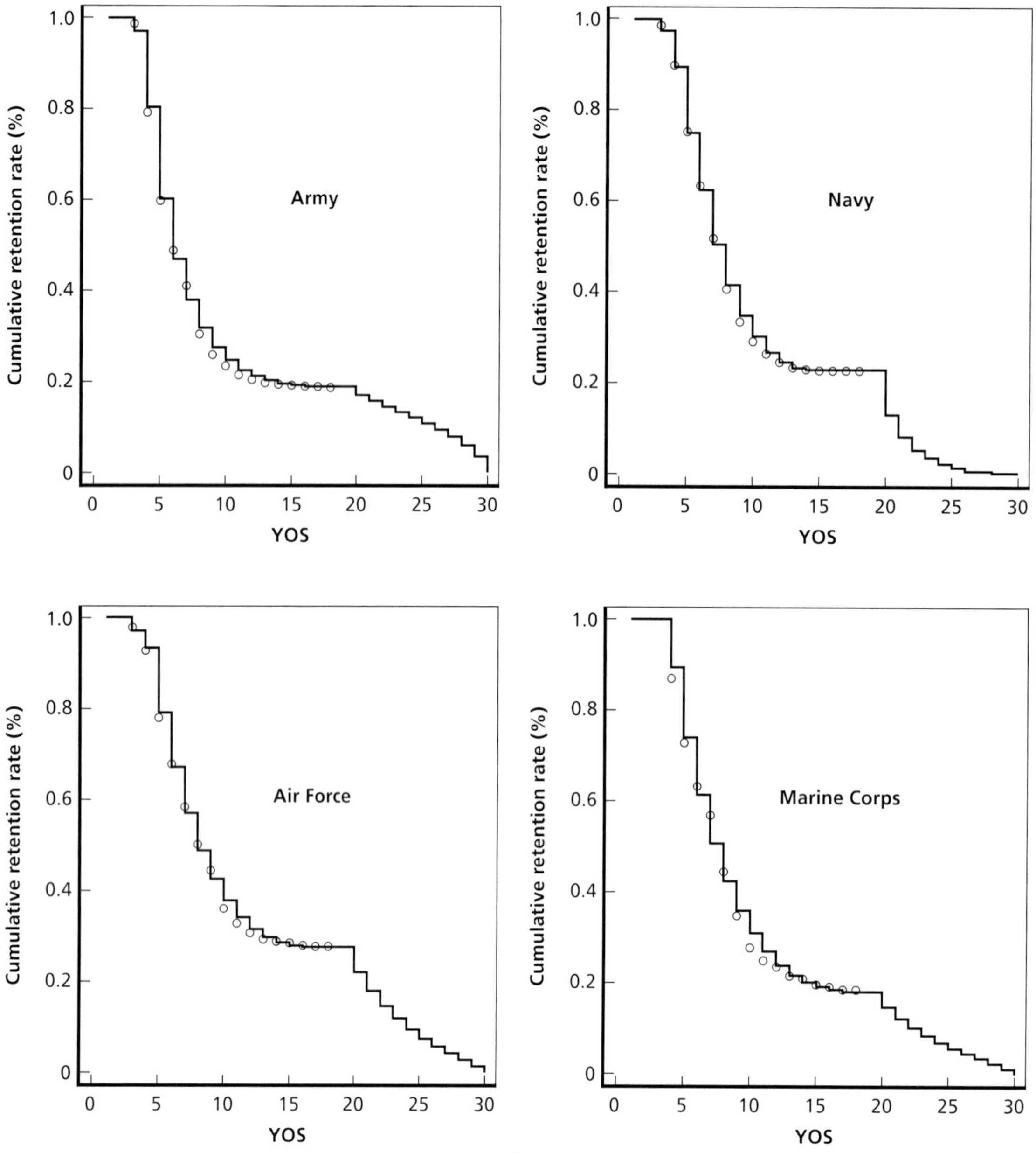

NOTE: The small circles are actual data of cumulative office retention to each year of service, and the lines are predicted retention.

RAND *TR1260-2.3*

CHAPTER THREE

Exploring Policy Options

This chapter builds on the modeling discussed in the previous chapter to explore policy options aimed at changing officer retention. The purpose of the analysis is to demonstrate the capability of our model to assess the retention effect of alternative policies rather than to assess a specific policy or recommend a particular policy action. The policies we consider are ones intended to influence the retention of officers in their midcareer.

The chapter begins with a description of the simulation capability that we built and demonstrates it for the Army by considering the retention effects of a bonus and of a separation pay scheme. As another demonstration, we then consider the retention effect of continuation pay within the context of multiyear contracts. In this policy exploration, we specifically consider the ACP program. The final policy we explore is one that provides a benefit to officers but requires that the officers incur an additional service obligation. In the case we consider, the benefit is the opportunity (and payment of costs) to attend graduate school. Specifically, we assess the retention effects of the GRADSO program. This third policy exploration is similar to the second one to the extent that each involves a benefit associated with a multiyear service obligation, but, as described later in the chapter, the third policy is more complex.

Simulation Capability Development and Demonstration

To conduct simulations of different policy options, we needed to develop a simulation capability. This capability inputs the parameter estimates and the DRM equations and predicts the resulting retention patterns, given the military and civilian pay lines. We developed the simulation capability in Microsoft Excel, using Visual Basic for Applications (VBA), because of Excel's wide use by military compensation analysts. The appendix describes the capability in more detail and provides users information about the formats of the spreadsheets, the inputs and outputs, and what information is predicted.

Figure 3.1 shows the baseline simulation for Army officers. The top two charts in the figure show the simulated year-to-year retention and cumulative retention rates by source of commission (nonscholarship ROTC, scholarship ROTC, and academy). The bottom two charts show the year-to-year and cumulative retention rates overall, which can be thought of as the appropriately weighted sum of the curves in the top of the figure.

The simulated retention profiles shown in the figure conform to expectation and what has been found in past research. Cumulative retention rates are near 100 percent in the initial years, when officers are under an initial service obligation, but begin to fall after three years (or later, depending on source of commission) once their initial obligation is complete and they

Figure 3.1
Baseline Simulations for the Army

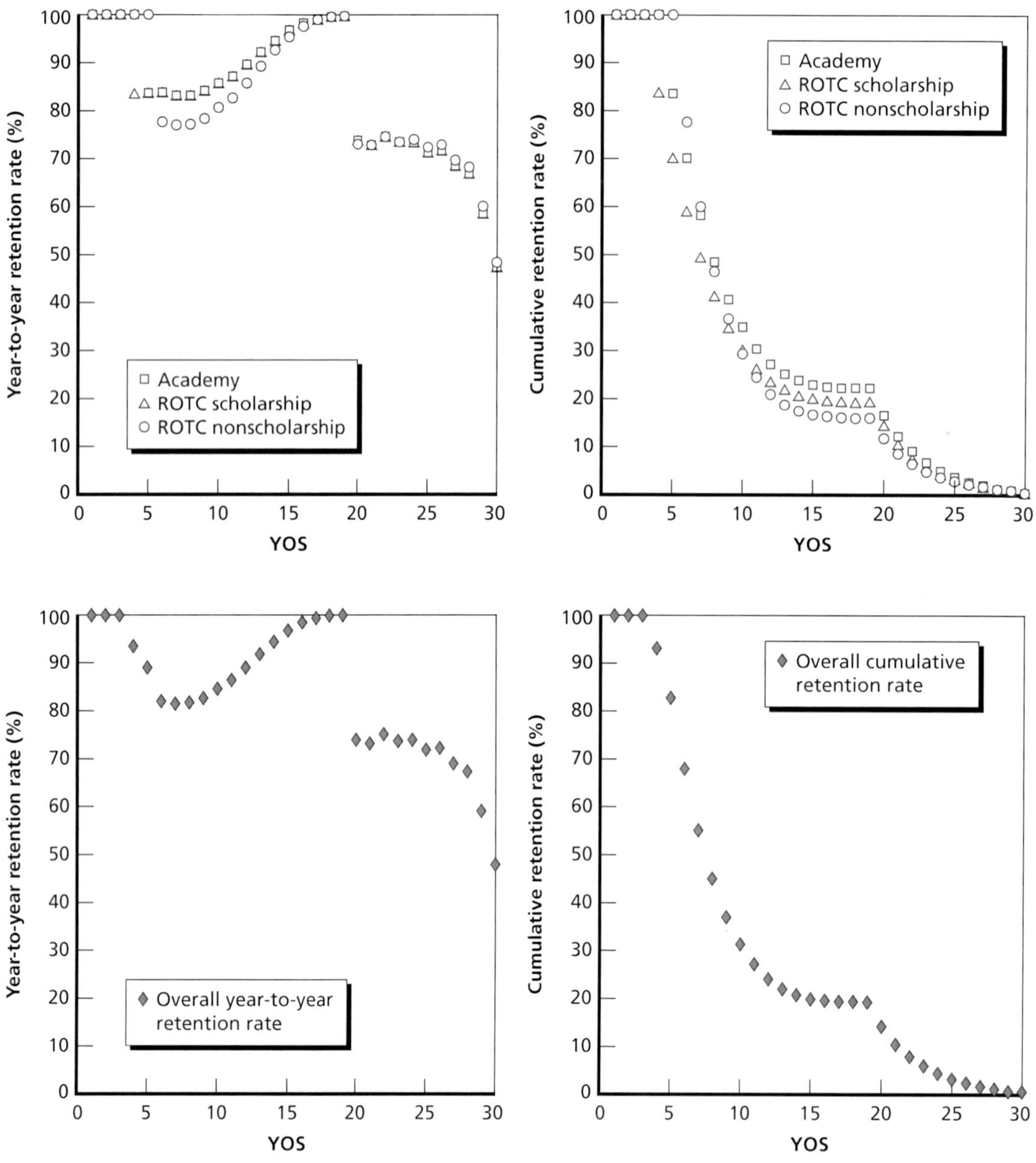

RAND *TR1260-3.1*

are free to leave service. Retention rates climb after ten years of service and stabilize after about 15 years because of the military retirement system. This system cliff vests members at 20 years of service in an immediate annuity that, under the so-called high-three system, pays about 50 percent of the average of the highest three years of basic pay. Once members are vested at 20 years, retention rates decline dramatically as members retire from service.

It is important to bear in mind that these curves show the steady-state results under existing military compensation and personnel policy. The cumulative retention rate can be regarded as either the population distribution across years of service given that accessions, personnel management policy, and compensation are constant, or as the cumulative retention rate for members of a particular cohort who experience the existing, rather than alternative, personnel management and compensation structure throughout their careers.[1] Because they show simulated retention patterns under existing policy, we call these results the *baseline* simulation results.

We can use this baseline simulation to explore the effect of changes in policy. We consider two demonstration policy excursions. First, we consider the effects that changes in bonuses and separation pays can have on the retention of midcareer officers via some specific numerical examples that are illustrative of the general effects of bonus pays and separation pays. We then consider the effects of a 10-percent increase in officer RMC. The DRM approach can also accommodate simulations of other types of compensation actions, including changes in the structure of the retirement system and changes in other special incentive pays (as is shown later in this chapter). Here, we demonstrate the capability for officers by considering the effects on retention of bonuses and separation pay and then of a change in RMC. We also note that the simulations can be extended to compute the cost of different policies, though we do not incorporate that capability here.

Change in Bonus and Separation Pay

We first examine the effect of a bonus pay offered in a single YOS, then the effect of a separation pay offered in a single YOS, then the combined effect of a bonus pay and a separation pay. These particular examples are all designed to result in an increase of 10 percent in the number of officers retained to the tenth year of service; this allows for some interesting comparisons across the examples and helps to illustrate how both bonuses and separation pays can be used to increase retention at a targeted year of service.

The steady-state effect of a bonus pay given at a particular YOS is to increase retention in the YOS in which the bonus is offered and to increase retention rates in those years preceding and following the bonus. Despite the fact that the bonus is offered in only one year, it has effects that propagate throughout the officer career cycle. This is illustrated by Figure 3.2, which shows the simulated effect of a $20,900 bonus in YOS 10, which we chose to result in a 10-percent increase in retention at the tenth YOS. The top two charts in the figure show the overall year-to-year and cumulative retention rates for both the baseline and the bonus policy. The bottom chart in the figure shows the relative change in the cumulative retention rate. The baseline and bonus policy curves in the top two charts show a just distinguishable difference, while the bottom chart, showing the relative change, shows the effect much more clearly. We see from the bottom chart that the effect of the bonus pay runs throughout the officer career rather than simply dampening out after a few years. Although the effect is clearly greatest at the targeted year of service, the single bonus results in a significant effect from YOS 7 onward. The increased retention before YOS 10 is attributable to officers looking forward to the bonus that will be available if they stay until YOS 10, and the increased retention thereafter is attributable to officers deciding to stay at a greater rate because of the pull of retirement and other benefits of staying in the military given that they have stayed until at least YOS 10, despite the

[1] A logical extension of the current model would be to integrate the DRM with an inventory model of personnel, which would yield predicted numbers that would reflect the current YOS structure of the force, as well as anticipated future accession policy.

Figure 3.2
Simulated Effect of a $20,900 Bonus in the Tenth Year of Service for the Army

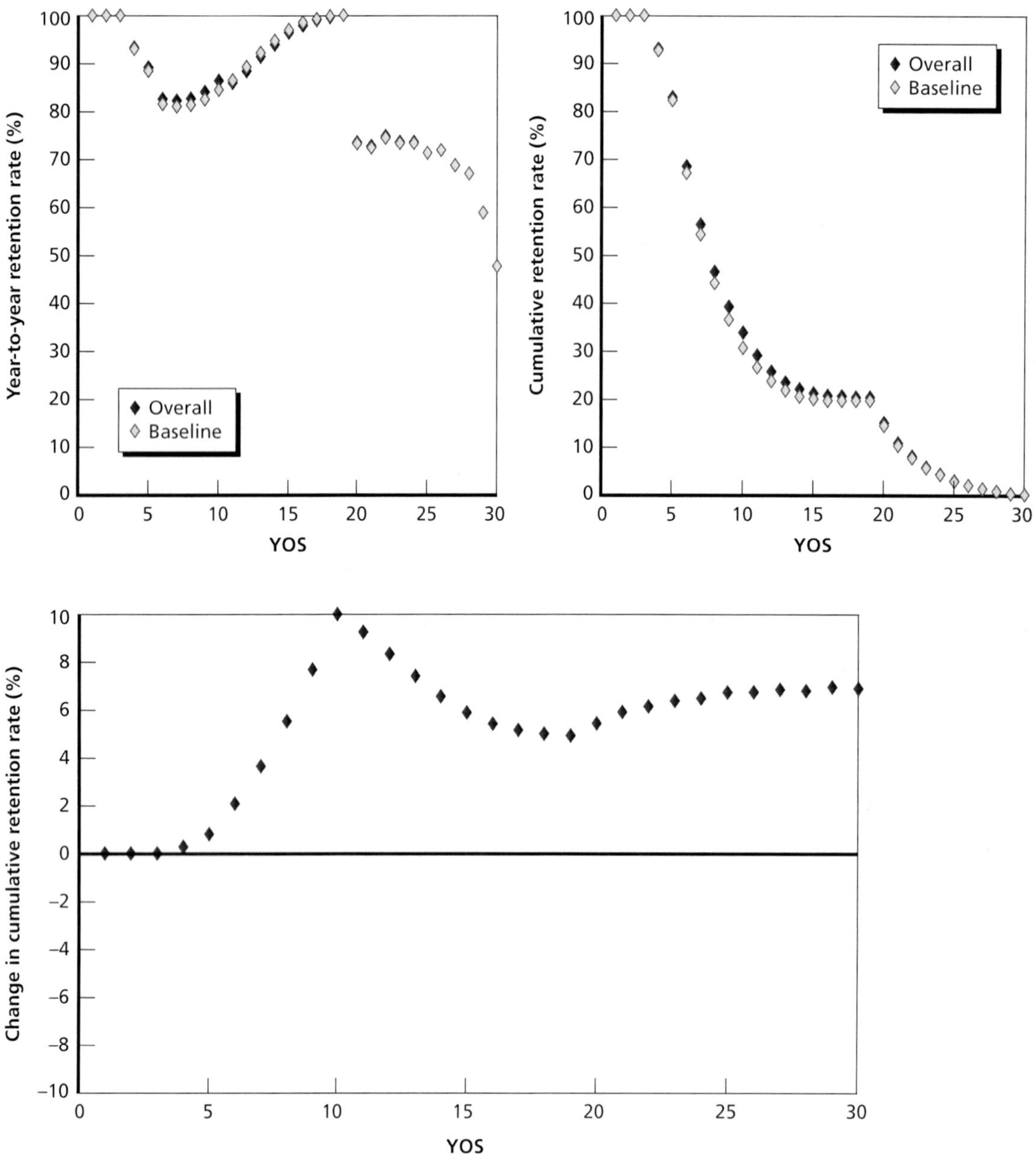

RAND *TR1260-3.2*

fact that the mean taste of the population of officers staying until and after YOS 10 is lower than in the baseline case.

In contrast, the steady-state effect of a separation pay offered in a particular YOS is not just to lower the cumulative retention rate in the year in which the separation pay is offered and for following years but also to *raise* the cumulative and year-to-year retention rates for those years preceding the year the separation pay is offered. Officers are pulled forward by the prospect of a separation pay being offered in a particular year. This can be seen in Figure 3.3, which shows the simulated effect of a $103,000 separation pay in YOS 11, which we chose to result in

Figure 3.3
Simulated Effect of a $103,000 Separation Pay in Year 11 for the Army

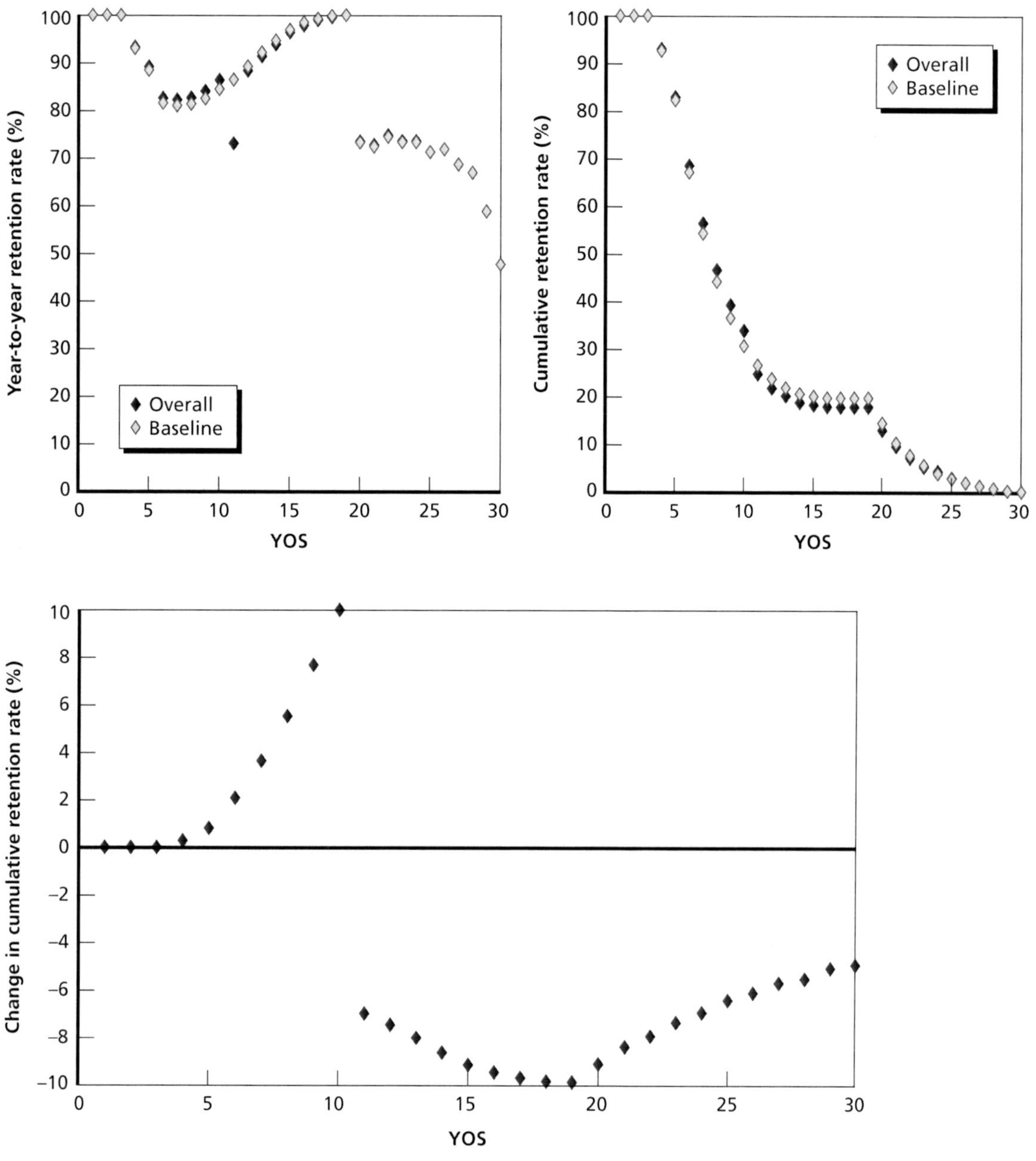

RAND *TR1260-3.3*

a 10-percent increase in retention at the tenth YOS for comparability with the first example. In addition, the change in cumulative retention rates before the tenth YOS is essentially identical with the change observed in the first bonus example. Cumulative retention falls in the year of the separation pay and for the following years, eventually resulting in 10 percent fewer officers reaching eligibility for retirement. As in the first example, significant effects can be seen throughout the officer career cycle, from YOS 7 on. The reason for the lower retention after YOS 11 involves the selection effect of this policy. The separation bonus increases the retention to YOS 10 of officers who otherwise would have left the military earlier. The retention of these officers would tend to decrease the average taste of officers completing YOS 10 relative to the baseline. At YOS 11, some officers leave. The stay-or-leave decision depends not only on taste but also on the shock, and some officers with a high taste for the military will draw a negative shock and leave. It appears that, on net, the officers who do not leave at YOS 11 have an average lower taste for the military than do the officers at that point in the baseline. As a result, retention is lower from YOS 12 to 20 than at baseline.

As another example, we show the combined effect of a bonus pay offered in one year with a separation pay offered in the following year. As in the first two examples, the effect of both the bonus pay and the separation pay is to increase retention in the years preceding the special pays. This can be seen in Figure 3.4, which shows the effect of a $14,400 bonus pay offered in the tenth YOS combined with the effect of a $39,300 separation pay offered in the 11th YOS. The bonus and separation pay amounts were selected to result in both an increase in the number of officers retained in the tenth YOS by 10 percent and so that the number of officers reaching eligibility for retirement would be identical to the baseline. As in the first two examples, the effect of the special pays reaches throughout the officer career cycle.

It can be seen from these basic examples that there is substantial flexibility in using simple bonuses and separation pays to shape the force,[2] and the model we propose can be used to examine the steady-state effect of bonuses and separation pays. Although we recognize that there are limitations to the steady-state approach, such a model can provide useful insights into the behavioral responses of officers to bonus and separation pays.

A Ten-Percent Increase in Officer Regular Military Compensation

We next demonstrate the model capability by showing the retention effects of a 10-percent increase in RMC. We would expect a pay increase to increase retention not only because current pay increases but also because future pay increases. And, because the DRM is a forward-looking model, officers are modeled to incorporate the change in current and future RMC into their current retention decision.

Figure 3.5 shows the simulated change in year-to-year retention rates and in the cumulative retention rate. The increase in the year-to-year retention rates is particularly noticeable in the midcareer, between YOS 6 and YOS 13. This is the period when officers make decisions about whether to stay for a full 20-year career. The simulations predict that a 10-percent increase in RMC substantially changes the likelihood that Army officers would stay in the midcareer.

In the next subsections, we consider more-complex compensation schemes, such as bonuses (or some other benefit) that are contingent on incurring an additional service obligation. Both alternatives have been implemented in the recent past, first the Air Force ACP program and then the Army GRADSO program.

The Air Force Aviator Continuation Pay Program

In this section, we explore the retention effects of a bonus program that also requires that the officer incur a multiyear service obligation. Specifically, we consider the Air Force ACP

[2] The bonuses and separation pays in the examples are simple in the sense that they do not require the officer to make any commitments, e.g., to stay a certain length of time or to leave at a certain future year of service.

Figure 3.4
Simulated Effect of a $14,400 Bonus in Year 10 Combined with a $39,300 Separation Pay in Year 11 for the Army

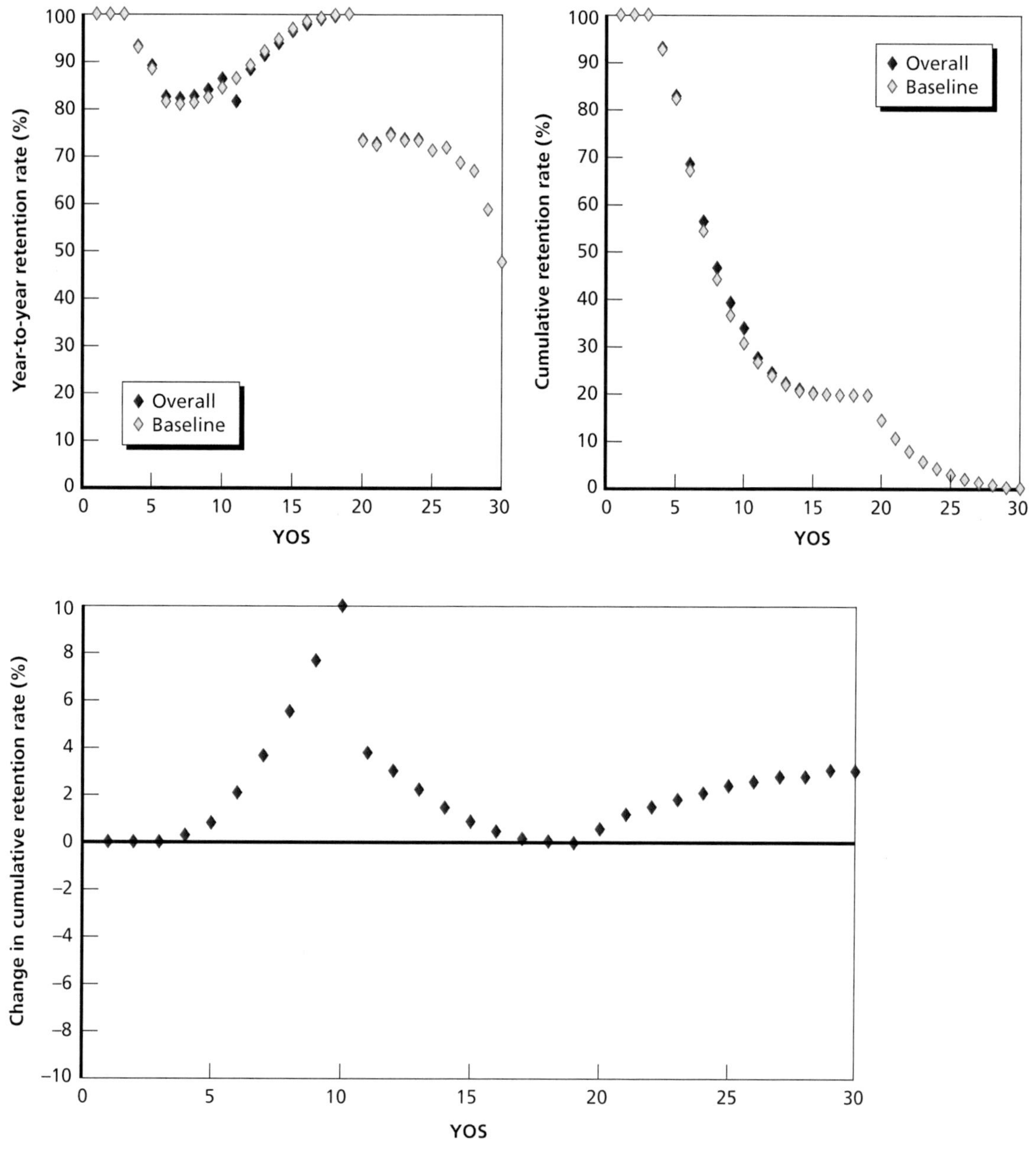

RAND *TR1260-3.4*

program. A multiyear agreement is a potentially useful retention tool because it provides the services with more certainty regarding the near-future retention rates for those occupations in which some fraction of officers are under a multiyear obligation. Under particular circumstances, it may also be more cost-effective than simple bonus pays in retaining officers to a particular YOS.

As noted in the previous chapter, we exclude pilots when we estimate the Air Force officer model. To demonstrate the capability of the model to handle benefits with multiyear service

Figure 3.5
Simulated Effect of a Ten-Percent Increase in Regular Military Compensation for Army Officers

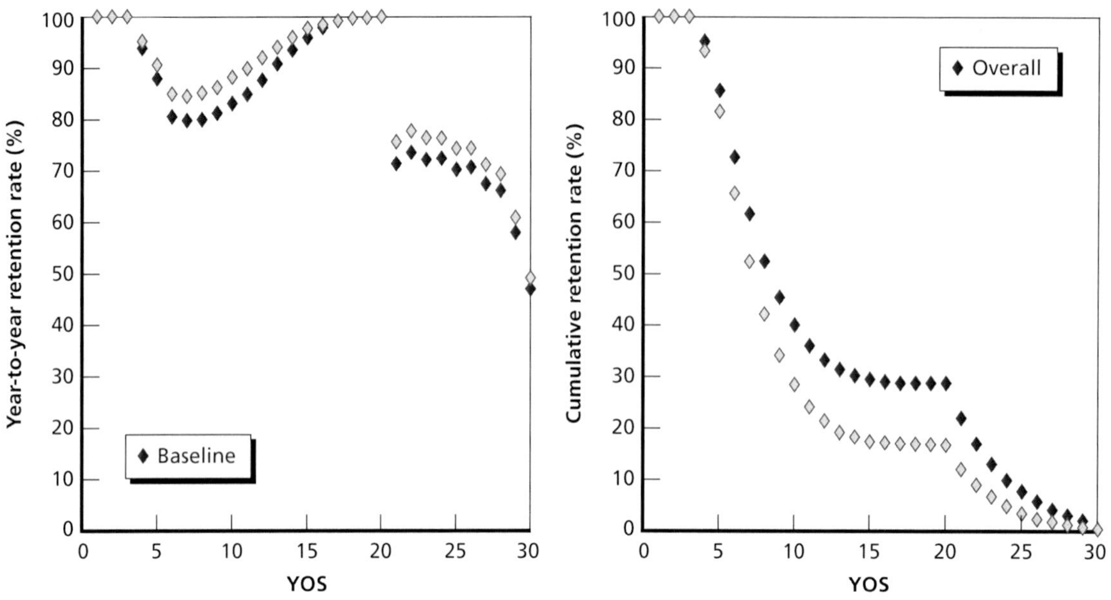

RAND *TR1260-3.5*

obligations and specifically the Air Force ACP program, we calibrate the parameters of the DRM to recent retention rates for pilots rather than estimate them. This involves selecting a set of parameters that most closely yield simulated retention rates that mimic actual retention rates in recent years. We discuss our modeling approach in more detail later in this subsection, but first we provide some background on the ACP program.

Background

In the early 1990s, the Air Force developed an incentive pay program to induce pilots to remain in service. The ACP program paid an annual bonus to pilots who committed to certain terms of service. Subsequently, the program was expanded to include not only pilots but also certain groups of navigators and air battle managers. The program has been revised from year to year in response to Air Force requirements and, more importantly, to changes in the outside civilian opportunities for rated officers (U.S. Air Force, 2009). We discuss later in this section the difficulty this uncertainty poses for estimating the DRM for pilots.

Model Details

This version of the DRM for the ACP models the program offered to pilots in 2008, in which only pilots who had just completed their training service obligation were allowed to opt in. To extend the model for ACP, we need to add another equation to compute the value of accepting the ACP contract. Under the ACP contract, the officer agrees to stay for five years, receiving a $25,000 bonus in each year. At the end of the ACP contract, the officer serves from year to year as under the basic DRM.

We add one equation to our two-equation model to express the value of the ACP program (see Table 3.1 for an explanation of the symbols in this model). The value of the ACP program to a particular officer consists of the present value of the officer's individual taste for the mili-

Table 3.1
Explanation of Symbols in the Aviator Continuation Pay Model

Symbol	Explanation
V_t^S	Value of staying for one more year
V_t^L	Value of leaving
V_t^{ACP}	Value of taking the 5-year ACP contract
β	Discount factor
γ^M	Taste preference for military service
W_t^M	Military wage
W^{ACP}	ACP bonus, currently $25,000 per year
W_t^C	Civilian wage, inclusive of retirement benefits
R_t	PDV of the military retirement benefit payable to officers retiring in year *t*
ε_t	Environmental disturbance at time *t*

tary, military wage, and ACP bonus pay over five years, plus the expected value of the maximum of staying or leaving after five years, plus the current environmental disturbance term:

$$V_t^S = \gamma^M + W_t^M + \beta E_t\left[\max\left(V_{t+1}^S, V_{t+1}^L\right)\right] + \varepsilon_t, \quad 3.1$$

$$V_t^L = R_t + \sum_{\tau=t}^{T} \beta^{\tau-t} W_\tau^C, \quad 3.2$$

and

$$V_t^{ACP} = \sum_{\tau=0}^{4} \beta^\tau \left(\gamma^M + W_\tau^M + W^{ACP}\right) + \beta^5 E_t\left[\max\left(V_{t+5}^S, V_{t+5}^L\right)\right] + \varepsilon_t. \quad 3.3$$

An eligible officer compares the value of leaving, V_t^L, with the maximum of the value of staying for one year, V_t^S, and the value of staying for five years and collecting the ACP bonus for each of the five years, V_t^{ACP}. Thus, the probability that an initially eligible officer will stay is

$$\Pr\left(\max\left(V_t^S, V_t^{ACP}\right) > V_t^L\right) = \Phi\left(\frac{\max\left(V_t^S, V_t^{ACP}\right) - V_t^L}{\sigma}\right).$$

Officers who are not eligible compare the value of staying for one year with the value of leaving, as in the basic model:

$$\Pr\left(V_t^S > V_t^L\right) = \Phi\left(\frac{V_t^S - V_t^L}{\sigma}\right).$$

Simulation Results for the Aviator Continuation Pay Model

Estimation of the DRM for pilots using data from the past decade is difficult because of the frequent changes in the details of the ACP program in recent years. Because of these frequent changes, our model would need to incorporate officer expectations about possible future changes in the program. This is relevant if the officer is allowed to sign up for the ACP either at any time during a multiyear window or at a particular future period. Either way, if there is uncertainty about the future availability and amount of the ACP, that uncertainty affects the expected value of the maximum of the choice between taking the ACP or not in the future period, and this feeds back to current decisions regarding staying or leaving. The effect on the expected maximum follows the same logic as the role of uncertainty from environment disturbances in the basic model (see the equation for expected value of the maximum and the surrounding discussion in the previous chapter). Mattock and Arkes (2007) has estimates using data from a relatively stable period for the ACP.

Here, we conduct simulations using parameters that were obtained by calibrating the DRM to recent retention rates for pilots. Calibration has the advantage that it is relatively quick and easy to do and provides a good "first pass" at a set of parameter estimates. The disadvantage is that the choice is more arbitrary than those obtained by estimating the model with data. Thus, the ideal approach for conducting simulations is to use parameters estimated from the data. Our objective here is to illustrate the simulation capability and provide an indication of the effects that ACP can have on pilot retention. Thus, as a first pass, calibration is adequate. The simulations focus on how changes to ACP bonus amount and contract duration could change retention rates and take rates for ACP contracts.

Figure 3.6 shows the simulated effect that a change in the annual bonus can have on cumulative retention rates for officers commissioned from the U.S. Air Force Academy (USAFA). The change in retention from varying the bonus shows the expected effect: Higher annual bonuses result in higher cumulative retention rates. The effect is monotonic, that is, higher bonuses always result in a higher cumulative retention rate, other things held constant.

We also consider the effect on changing the duration of the service obligation of the ACP contract. An increase in duration by one results in another year of ACP bonus payments of $25,000 but also means another year of no flexibility to respond to an unexpected good civilian opportunity. Our simulation results are initially surprising, as seen in Figures 3.7 and 3.9. The effect of a change in contract duration is not monotonic. That is, lengthening the duration of the ACP contract does not always increase (or decrease) the cumulative retention rate (Figure 3.7) or the take rate (Figure 3.9). The three-year-duration contract results in a lower

Figure 3.6
Simulated Effect That Changes in Annual Aviator Continuation Pay Bonus Could Have on the Cumulative Retention Rate for U.S. Air Force Academy Commissions

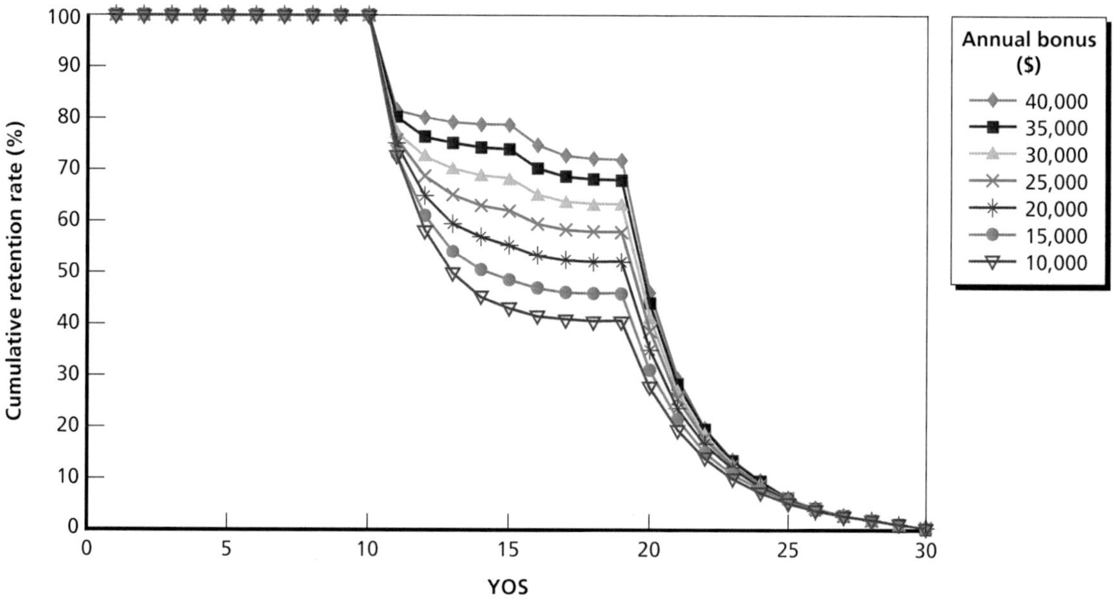

RAND *TR1260-3.6*

Figure 3.7
Simulated Effect That Changes in Aviator Continuation Pay Contract Duration Could Have on the Cumulative Retention Rate for U.S. Air Force Academy Commissions

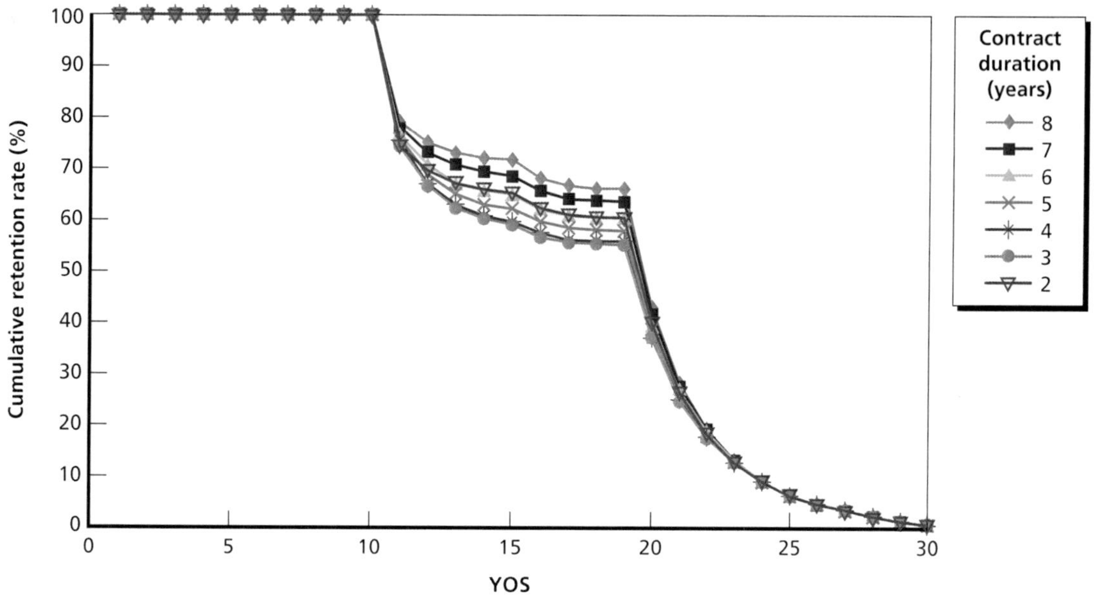

RAND *TR1260-3.7*

cumulative retention rate than the two-year contract, but, beyond three years, the retention rate is monotonically increasing in contract length. This inversion occurs because of the change in the value of the discounted stream of bonus payments relative to the value lost by committing to a certain service obligation. That is, the longer contract durations restrict individuals' flexibility to change their retention decisions as information about future uncertain events are revealed. The reduced flexibility has an adverse effect on retention of longer contract durations. This adverse-flexibility effect offsets the higher ACP bonus effect early on, between two and three years, but not later on, after contract lengths of four years. This suggests that the cost of reduced flexibility increases with contract length but at a decreasing rate. The value of military retirement benefits is important to the underlying calculation. In particular, the effect of military retirement benefits, which can be drawn after completing 20 years of service, grows larger the longer the contract and hence the fewer the years to qualify for retirement benefits.

These results can also be seen in the simulated changes to the overall take rates presented in Figures 3.8 and 3.9. Figure 3.8 shows the simulated change in the take rate in response to a change in the bonus offered, and, as expected, increasing the annual bonus increases the take rate at an ever-decreasing rate. Figure 3.8 shows that the take rate associated with a two-year ACP contract with a bonus of $25,000 per year is higher than that associated with a three-year contract, but, beyond three years, the simulation shows a consistent increase with contract length.

Although this program is effective in encouraging midcareer officers to stay, it also results in a more senior force overall. Once officers have served their obligation under ACP, they tend to be pulled forward by the prospect of the military retirement benefit.

Figure 3.8
Simulated Effect That Changing Annual Aviator Continuation Pay Bonus Could Have on the Overall Take Rate

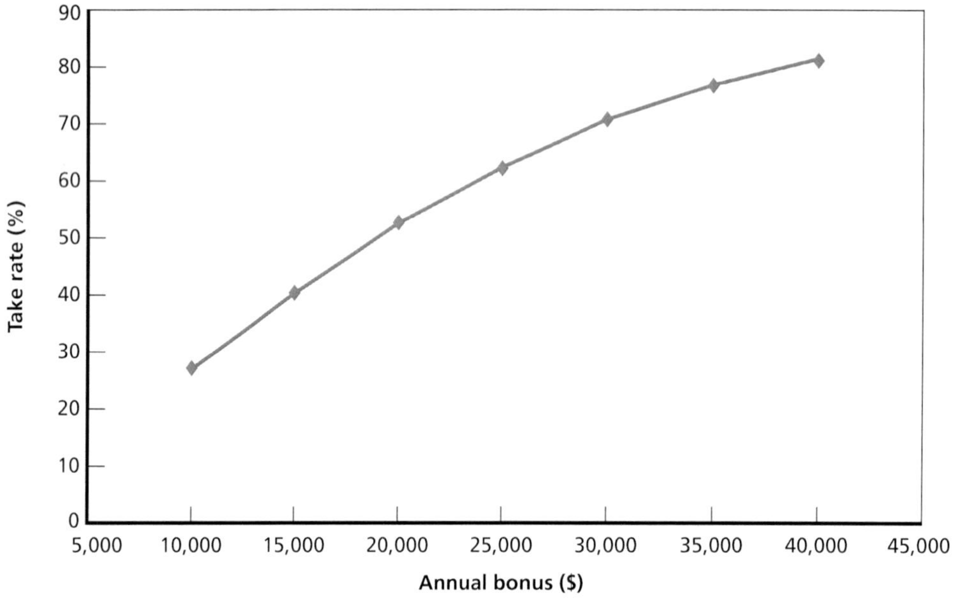

RAND *TR1260-3.8*

Figure 3.9
Simulated Effect of Changing the Aviator Continuation Pay Contract Duration on the Overall Take Rate

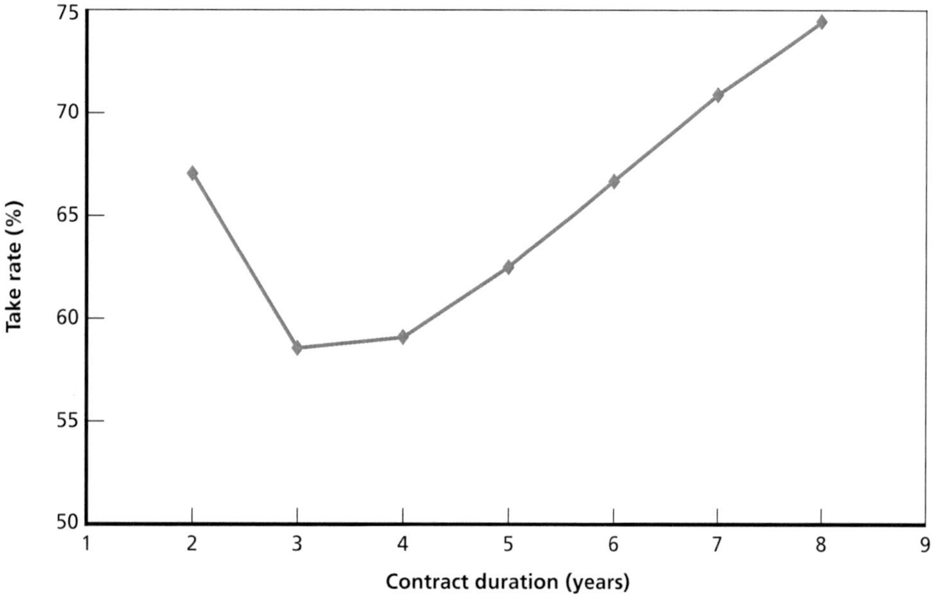

RAND *TR1260-3.9*

The Army Graduate School for Service Program

We next explore a policy alternative that provides a benefit with a multiyear service obligation, but the structure is more complex than the one considered in the previous section. Here, we consider the Army GRADSO program. The primary difference between GRADSO and ACP considered in the previous section is that the GRADSO service obligation is split into two parts: The first obligation provides an officer with the option, but not the requirement, to attend graduate school in exchange for a further service obligation of three years. It also contrasts with the Air Force example in that no benefit is paid during the additional three years of the initial service obligation, and a benefit is paid only if an officer decides to use (or exercise) the graduate school option after serving the full length of the extended initial service obligation. We begin this section with background on the GRADSO program, provide modeling details, and then present the simulation results.

Background

The Army's GRADSO program was initially approved by the Secretary of the Army in May 2005 and started with year group 2006 (that is, officers commissioned in 2006). The program offers the option to attend graduate school in exchange for a three-year obligation to be served consecutively with the commission service obligation. If the option to attend graduate school is exercised, the officer incurs an additional service obligation at a rate of 3:1 for time spent in graduate school. It appears that the program was more successful than initially anticipated, so it was capped at 300 participants as of year group 2009 (Campion, 2009).

Figure 3.10 shows the nominal timeline for the program. The initial extension to the commissioning obligation, called GRADSO, lasts three years. After GRADSO, the officer has the option to attend graduate school. If the officer does not attend graduate school, then he or she can either continue in service or separate. Officers who go to graduate school can incur a further obligation that carries them through their 14th to 16th year, depending on the length of their initial commission service obligation plus the three years from GRADSO.

GRADSO was offered in conjunction with two other programs, BRADSO, which offers a service branch of choice for officers who would otherwise not be chosen because of their class rank in exchange for an additional service obligation of three years, and PADSO, which offers an initial post of choice in exchange for an additional service obligation of three years. Officers were able to select to participate in at most two of the three programs, and the additional service obligations were to be served consecutively. In the model and simulation example below, we consider only GRADSO.

Model Details

We extended the model to reflect the expanded choices available. The first choice is whether or not to participate in GRADSO. We make the simplifying assumption that the decision to participate in GRADSO is made at the end of the initial commissioning service obligation; in actuality, the decision is made at the time of commissioning. The second choice, for an officer deciding to participate in GRADSO and successfully completing the service obligation, is whether or not to go to graduate school. We make the simplifying assumption that officers can choose either to go to graduate school for two years or not at all; officers actually can choose any graduate school length up to two years.

This means that we need to add two new equations to our model (see Table 3.2 for an explanation of the symbols in this model): one to reflect the value of taking GRADSO and thus obtaining the option to attend graduate school, and one to show the value of attending graduate school given that the option is available to the officer:

$$V_t^S = \gamma^M + W_t^M + \beta E_t\left[\max\left(V_{t+1}^S, V_{t+1}^L\right)\right] + \varepsilon_t, \tag{3.4}$$

$$V_t^L = R_t + \sum_{\tau=t}^{T} \beta^{\tau-t} W_\tau^C, \tag{3.5}$$

Figure 3.10
Army Graduate School for Service Program Timeline

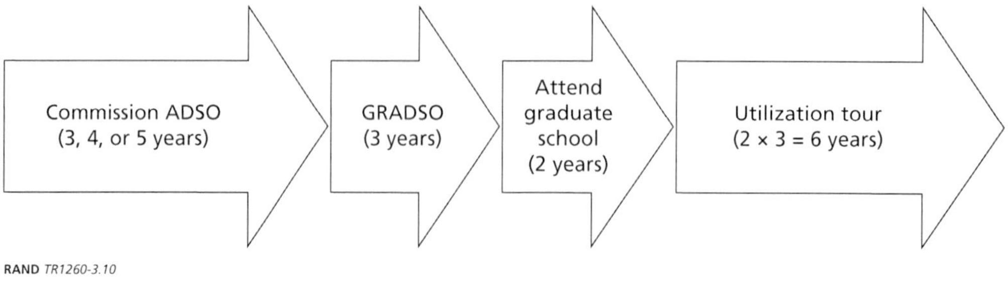

RAND *TR1260-3.10*

Table 3.2
Explanation of Symbols in the Army Graduate School for Service Model

Symbol	Explanation
V_t^S	Value of staying for one more year
V_t^L	Value of leaving
V_t^G	Value of incurring GRADSO and the option to attend graduate school
V_t^D	Value of attending graduate school and incurring the additional service obligation
β	Discount factor
γ^M	Taste preference for military service
γ^D	Taste preference for attending graduate school and the PDV of the change in lifetime earnings
W_t^M	Military wage
W_t^C	Civilian wage, inclusive of retirement benefits
R_t	PDV of the military retirement benefit payable to an officer retiring in year *t*
ε_t	Environmental disturbance at time *t*

$$V_t^G = \sum_{\tau=0}^{2} \beta^\tau \left(\gamma^M + W_\tau^M\right) + \beta^3 E_t\left[\max\left(V_{t+3}^S, V_{t+3}^D, V_{t+3}^L\right)\right] + \varepsilon_t, \tag{3.6}$$

and

$$V_t^D = \gamma^D + \sum_{\tau=0}^{7} \beta^\tau \left(\gamma^M + W_\tau^M\right) + \beta^8 E_t\left[\max\left(V_{t+9}^S, V_{t+9}^L\right)\right] + \varepsilon_t. \tag{3.7}$$

An officer compares the value of leaving, V_t^L, with the maximum of the value of staying for one year, V_t^S, and the value of staying for three years and obtaining the option to attend graduate school, V_t^G. Thus, the probability that an officer will stay is

$$\Pr\left(\max\left(V_t^S, V_t^G\right) > V_t^L\right) = \Phi\left(\frac{\max\left(V_t^S, V_t^G\right) - V_t^L}{\sigma}\right).$$

If an officer has chosen to take GRADSO, immediately upon completion of his or her extended obligation, the officer compares the value of leaving, V_t^L, with the maximum of the value of staying for one year, V_t^S, and the value, V_t^D, of exercising the option, which, as mentioned, means two years of graduate school plus six additional years of service for a total of eight additional years. The present value of graduate school is given by the γ^D term, which we vary in our simulation below. The use of a single term γ^D is a simplification because graduate school might add to or subtract from military taste while attending graduate school as a service member, in subsequent years of military service, and in civilian jobs. Further, because we noted above that the taste term can also be thought of as capturing idiosyncratic components of military and civilian earnings, it is also possible for graduate school to alter these components. A full structural model would include terms to account for all of these possible effects, and the use of a single term to represent the present value of graduate school implicitly assumes that graduate school adds a constant amount in each period regardless of whether one is in the military or not. Thus, the probability that an officer will stay after completion of GRADSO is

$$\Pr\left(\max\left(V_t^S, V_t^D\right) > V_t^L\right) = \Phi\left(\frac{\max\left(V_t^S, V_t^D\right) - V_t^L}{\sigma}\right).$$

Officers who decline to participate in GRADSO, who decline to go to graduate school, or who have completed their graduate school obligation simply compare the value of staying for one year with the value of leaving, as in the basic model:

$$\Pr\left(V_t^S > V_t^L\right) = \Phi\left(\frac{V_t^S - V_t^L}{\sigma}\right).$$

Simulation Results of the Army's Graduate School for Service Program, Without Cap

The simulations of the effects on retention of the Army's GRADSO program used the Army parameters shown in Table 2.2 in Chapter Two. Figure 3.11 shows how the steady-state simulated cumulative retention rate varies as with the assumed discounted present value of attending graduate school. This is what we would have expected to see if the original program offered to year groups 2006, 2007, and 2008 had continued in perpetuity. What is remarkable is the relatively large effect of even a relatively small assumed value for graduate school. We note that this is the value of graduate school to the individual. It does not reflect the costs to the Army of the GRADSO program and, in particular, does not reflect the opportunity cost associated with officers not being available while they are in graduate school, nor the increased cost associated with a higher proportion of officers reaching retirement eligibility. But it does seem safe to say that a delayed benefit associated with a service obligation shows some promise as a cost-effective means of inducing officers to extend their careers in the military.

The model seems to perform reasonably well compared with the Army's projections of Army officer retention, which indicate an increase of 19 percentage points over baseline of officers reaching eight YOSs for year groups 2007 and 2008.[3] From Figure 3.11, we see that this

[3] As reported in Campion (2009, p. 13).

Figure 3.11
Simulated Effect of the Army Graduate School for Service Program with No Cap on the Number of Participants, Varying Present Value of Graduate School

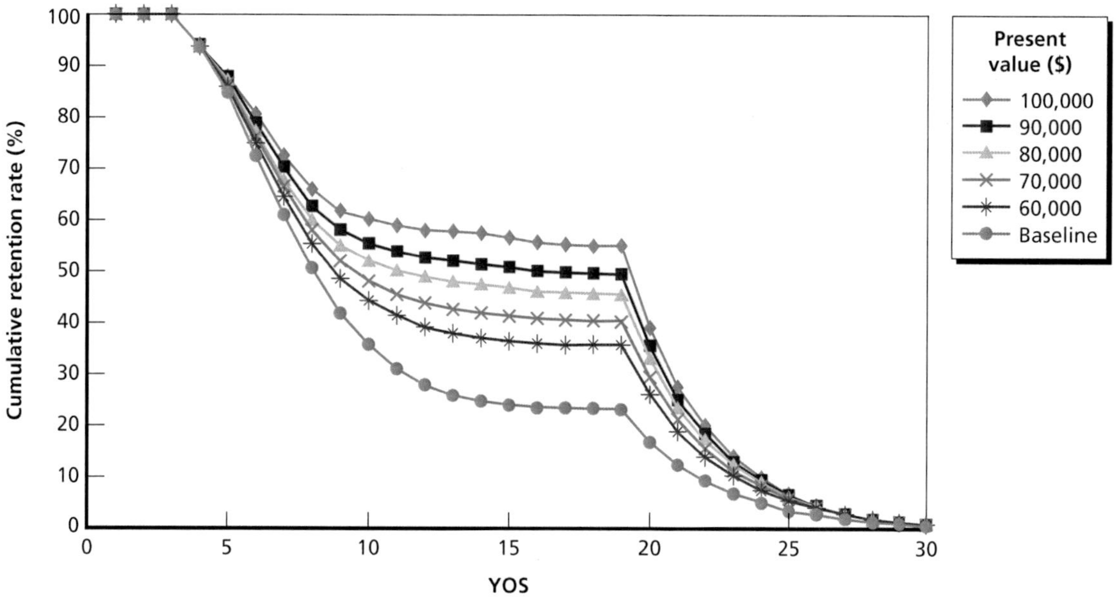

RAND *TR1260-3.11*

percentage-point increase is attained when the presumed value of graduate school to the individual is about $100,000. In general, a large element of the cost of college or graduate school is forgone earnings, but an Army officer's salary continues to be paid while the officer attends graduate school and so there are no forgone earnings. Further, it is unlikely that the graduate schooling will have much effect on the officer's earnings while he or she is in the military, and officers who accept GRADSO and attend school will most likely stay to 20 years of service or more, as Figure 3.11 indicates. This means that the likely positive effect that graduate school has on earnings will be mainly through civilian earnings, and this effect will not be realized until, say, five or more years after completing graduate school. It seems unlikely, however, that the civilian earnings increase would be large enough to account for the perceived value ($100,000) of two years of graduate education, which suggests that a significant portion of the value is nonpecuniary.[4]

One difficulty inherent in this type of program is the in-kind nature of the benefit, or, stated differently, the value of the benefit to the officer is not under the Army's control. This makes it difficult to adjust the benefit in response to new information on the propensity of officers to take the benefit. When more officers took advantage of this program than seemed appropriate for the needs of the Army, the only choice available was to cap the number of participants in the program. In cases in which it is important for policymakers to have fine control

[4] If an officer worked in the civilian world for 25 years after leaving the military at YOS 23 and had a personal discount rate of 10 percent, a civilian wage increment of roughly $42,000 per year would be needed to produce a present value of $100,000 at YOS 8. If one year of college increases wages by 10 percent, then baseline civilian earnings of nearly $210,000 are needed $(210,000 \times 2 \times 0.10 = 42,000)$. This is high in view of the average weekly earnings of, say, full-time, full-year white males of ages 32–36 in professional technical occupations, which, in 2009, were $1,680 per week or $87,300 for a 52-week year (authors' tabulations with CPS data), so it seems likely that much of the perceived worth of graduate education is nonpecuniary.

over the value of the benefit, cash benefits are the obvious choice. Still, in-kind benefits, such as the option to attend graduate school, may exert a powerful draw on officers a service would like to keep and might, or might not, be worth the cost borne by the service. The results show that in-kind benefits and cash bonuses are both effective tools for shaping the force.

Another difficulty inherent in this type of program comes from the long-term nature of the agreements. The Army has made a commitment to support graduate school for a sizable number of officers in year groups 2006–2008, and the consequences of this commitment will play out for more than a decade as a large "pulse" of officers flows through the system. Thus, policymakers may find it more useful to focus on arrangements that incur less enduring obligations on a service, such as contracts of shorter duration.

CHAPTER FOUR

Concluding Thoughts

To manage officer retention, policymakers require a capability that will enable them to conduct what-if exercises of different policy alternatives. Such a capability is needed by service, as well as for key officer communities. This study built on past efforts for both officers and enlisted personnel to develop a DRM of officer retention behavior that provides such a capability. We estimate this model for all four services and, using the estimates for the Army, build a spreadsheet-based simulation capability that allows policymakers to conduct what-if analyses. The report summarizes the analyses and illustrates different policy options.

The policy simulations illustrate how the effect of a single bonus or separation pay reverberates throughout officer careers. Rather than rapidly diminishing over time, effects can propagate ten or 20 years in the future. This happens because bonuses and separation pays affect the distribution of officer preferences for those officers who choose to stay. Effects can also propagate to the beginning of an officer's career because bonus and separation pays pull officers forward who would otherwise leave.

The policy simulations also show by example that the same retention target for a given year can be met either through a single bonus or separation pay or via a combination of a bonus and a separation pay. The simulations further show how multiyear commitments associated with either a cash or in-kind benefit can exert a powerful influence on officer decisions and how bonuses in combination with a service obligation may serve as a more economical alternative to unconditional bonus pays. Finally, the simulations illustrate how an option to attend graduate school can improve midcareer officer retention, though perhaps can lead to a larger increase in officers continuing on to 20 or more years of service than a service prefers.

Recommendations for Further Work

The results of the analysis point to two veins of future work, one dealing with data refinement and one with further development of the model to make it more salient to analysts and policymakers.

The results of the statistical estimation procedure show that the feasibility of this approach depends critically on the quality of the data, particularly data salient to the initial obligation faced by an officer and whether a stay-or-leave decision was voluntary or involuntary. Ideally, DMDC will, in the future, collect such data from each of the services to facilitate statistical estimation of models of officer behavior. Meanwhile, calibration provides a useful, but not ideal, alternative approach.

The steady-state model we present can be used to gain useful insights into officer behavior over time. Future work that marries this model of officer behavior to an equilibrium model of promotion or of officer inventory, or both, would be a useful direction to pursue in future work. This would involve bringing promotion speed and up-or-out rules into the model, as has been done in earlier work, and also pairing the model with an inventory projection model such that there was interaction between retention and promotion (e.g., higher retention leading to slower promotion, leading to lower retention, and so forth). In addition, further work to go beyond "what if"—to come up with, say, the cost-minimizing bonus and separation pay schedule to reach a given YOS profile—could provide additional insights to military compensation analysts and policymakers. Also, work is under way at RAND to extend the approach to analyze not a steady state but the dynamic effects of introducing a policy affecting incumbent service members.

The extension of the model to include reserve participation is also desirable. In work under way, we are finding that the inclusion of reserve participation leads to an excellent fit with the data, precise parameter estimates, and an expanded policy analytic capability, namely, the capability to see how changes affecting either active or reserve compensation, or both, affect active component retention and reserve participation. Further refinement of the model and the possible development of a spreadsheet version of it await future work.

APPENDIX

Spreadsheet Implementation

The spreadsheet versions of the DRM are implemented using functions written in VBA, the macro language for Microsoft Excel. The VBA functions both read data stored in the spreadsheet (e.g., the bonus or separation pay for a particular year) and are called by cells in the spreadsheet to produce model results (e.g., the predicted retention rates for each year of service). The mathematical structure of the DRM is given, and the spreadsheet model can accommodate different empirical inputs, e.g., different parameter values, civilian earnings, military pay, or military retirement benefits. Thus, although we present a specific implementation here based on recent estimates and pay data, the user can potentially change or update the inputs.

The exposition of the spreadsheet and the associated VBA code is organized into five subsections. The first subsection explains the structure of the spreadsheets' worksheets, including the named ranges that are used in the VBA code.

The second subsection deals with code common to all versions of the spreadsheet DRM, mainly utility code for computing functions of the probability distributions used in the model and code for setting the expected civilian wage by year, the expected military wage by YOS, for reading values of the bonus pays and separation pays from the appropriate spreadsheet cells, and for calculating the discounted present value of military retirement benefits to which an officer is entitled having served until a given year.

The third subsection deals with the code used in the "basic" version of the DRM, in which, at a given YOS, officers face only the decision of staying for one more year or leaving in that year.

The fourth subsection explains the code for a version of the DRM that models the Air Force ACP program for pilots circa 2008, in which initially eligible officers could choose to sign a contract for an additional five-year obligation with a bonus pay of $25,000 per year, stay from year to year, or leave. (Once an Air Force officer taking ACP has served his or her five-year obligation, the officer also makes retention decisions from year to year.)

The fifth and final subsection examines the code for a version of the spreadsheet DRM that models the Army GRADSO program as initially implemented (i.e., no cap on the number of officers who could enter into the program; the current version of the program caps participation at 300 officers per year). Under the GRADSO program, an officer initially signs up for an additional three-year obligation beyond his or her initial commissioning obligation of three, four, or five years. Once the additional three years have been served, the Army officer has the option of attending graduate school for up to two years (with the Army paying all costs), incurring an additional obligation of three years for every year in graduate school.

Spreadsheet Structure

The spreadsheet workbooks consist of two worksheets ("SimpleDRM" and "Parameters") and a single module of VBA code for the DRM formulas. This structure is common to all versions of the spreadsheet DRM. The figures shown are for the basic DRM, but the structure of the worksheets is very similar across all the models. The substantive differences between models are solely in the VBA code particular to each model. This design feature is intentional because it is typically much easier to understand the difference between functions written in VBA than the difference between spreadsheets with different cell formulas because cell formulas are generally not directly visible and their operation can be obscured by reference to cell ranges that have no intrinsic meaning.

The spreadsheet also uses named ranges to refer to model parameter values or ranges where users will input values (such as bonuses or separation pays). Using named ranges rather than "raw" cell references makes the VBA code and cell formulas easier to understand. The named ranges are shown in Table A.1.

Figure A.1 shows the appearance of the first worksheet ("SimpleDRM") in the spreadsheet. To the left are two columns of highlighted cells that correspond to the ranges where users may enter the bonus pays or separation pays for a particular year. The rest of the cells show the model results, either cumulative retention rates or year-to-year retention rates by YOS for different source of commissions, as well as over all sources of commission. To the right are graphs showing either cumulative retention rates or year-to-year retention rates. Figure A.2 gives a zoomed view of columns A through K of the spreadsheet, showing YOS, bonus amounts, and the retention rates.

Model parameters are stored in the "Parameters" worksheet, shown in Figure A.3.

The spreadsheet invokes the DRM through calling the **Simulatel** function. This can be seen in Figures A.4 and A.5, where **Simulatel** appears in columns D, E, and F of the spreadsheet. **Simulatel** is an Excel array function; rather than being invoked cell by cell, it is instead called by and returns to a 30-by-1 range of cells. The **Simulatel** function takes as arguments the parameters of the model for the mode and scale of the taste distribution (**Alpha** and **Delta**

Table A.1
Named Ranges in the Spreadsheet Dynamic Retention Model

Name	Range
Beta	=Parameters!$E$2
Sigma	=Parameters!$E$3
Alpha	=Parameters!$E$4
Delta	=Parameters!$E$5
AlphaROTC	=Parameters!$E$6
DeltaROTC	=Parameters!$E$7
YOSRange	=SimpleDRM!$A$7:$A$36
WbRange	=SimpleDRM!$B$7:$B$36
WsRange	=SimpleDRM!$C$7:$C$36

Figure A.1
The SimpleDRM Worksheet

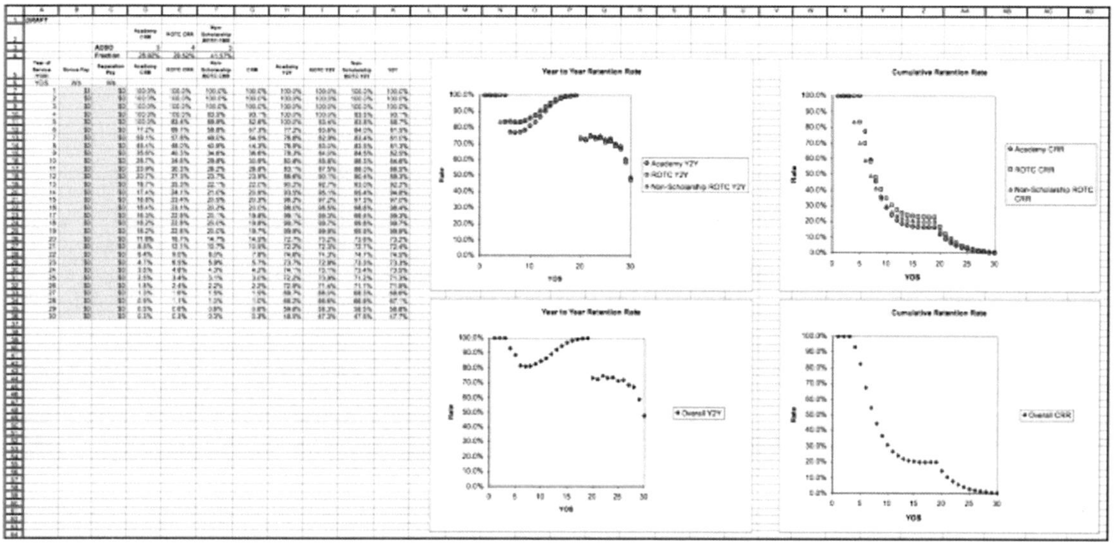

RAND *TR1260-A.1*

in the case of academy commissions, or **Alpha+AlphaROTC** and **Delta+DeltaROTC** in the case of ROTC commissions), the variance of the environmental disturbance term (**Sigma**), and the personal discount rate (**Beta**), as well as the active-duty service obligation for the particular source of commission. In addition, **Simulate1** takes as arguments the cell ranges corresponding to bonus pays (**WbRange**) and separation pays (**WsRange**); it does this so that Excel will automatically recompute the simulated retention rates if the values for either bonus pays or separation pays change. The remainder of the spreadsheet formulas consist of simple manipulations of the results returned by **Simulate1**, such as the weighted average to get the overall cumulative retention rate that appears in column G or the year-to-year retention rates that appear in columns H through K.

Common Utility Code

The code in this section is common to all versions of the spreadsheet DRM and consists mainly of utility code for computing functions of the probability distributions used in the model and code for setting the expected civilian wage by year and the expected military wage by YOS, for reading values of the bonus pays and separation pays from the appropriate spreadsheet cells, and for calculating the discounted present value of military retirement benefits to which an officer is entitled if he or she leaves in a given year of service.

Common Utility Code Listing

```
Option Explicit
Option Base 1
Public Const T As Integer = 31
Public Const yearMin As Integer = 4
Public Const yearMax As Integer = 30
```

Figure A.2
Columns A Through K of the SimpleDRM Worksheet

	A	B	C	D	E	F	G	H	I	J	K
1											
2				Academy CRR	ROTC CRR	Non-Scholarship ROTC CRR					
3			ADSO	5	4	3					
4			Fraction	28.92%	29.52%	41.57%					
5	Year of Service (YOS)	Bonus Pay	Separation Pay	Academy CRR	ROTC CRR	Non-Scholarship ROTC CRR	CRR	Academy Y2Y	ROTC Y2Y	Non-Scholarship ROTC Y2Y	Y2Y
6	YOS	Wb	Ws								
7	1	$1	$0	100.0%	100.0%	100.0%	100.0%	100.0%	100.0%	100.0%	100.0%
8	2	$0	$0	100.0%	100.0%	100.0%	100.0%	100.0%	100.0%	100.0%	100.0%
9	3	$0	$0	100.0%	100.0%	100.0%	100.0%	100.0%	100.0%	100.0%	100.0%
10	4	$0	$0	100.0%	100.0%	83.5%	93.1%	100.0%	100.0%	83.5%	93.1%
11	5	$0	$0	100.0%	83.4%	69.9%	82.6%	100.0%	83.4%	83.8%	88.7%
12	6	$0	$0	77.2%	69.7%	58.8%	67.3%	77.2%	83.6%	84.0%	81.5%
13	7	$0	$0	59.1%	57.8%	49.0%	54.5%	76.6%	82.9%	83.4%	81.0%
14	8	$0	$0	45.4%	48.0%	40.9%	44.3%	76.9%	83.0%	83.5%	81.3%
15	9	$0	$0	35.6%	40.3%	34.6%	36.6%	78.3%	84.0%	84.5%	82.5%
16	10	$0	$0	28.7%	34.6%	29.8%	30.9%	80.8%	85.8%	86.3%	84.6%
17	11	$0	$0	23.9%	30.3%	26.2%	26.8%	83.1%	87.5%	88.0%	86.5%
18	12	$0	$0	20.7%	27.3%	23.7%	23.9%	86.6%	90.1%	90.4%	89.3%
19	13	$0	$0	18.7%	25.3%	22.1%	22.0%	90.2%	92.7%	93.0%	92.2%
20	14	$0	$0	17.4%	24.1%	21.0%	20.9%	93.5%	95.1%	95.4%	94.8%
21	15	$0	$0	16.8%	23.4%	20.5%	20.3%	96.2%	97.2%	97.3%	97.0%
22	16	$0	$0	16.4%	23.1%	20.2%	20.0%	98.0%	98.5%	98.6%	98.4%
23	17	$0	$0	16.3%	22.9%	20.1%	19.8%	99.1%	99.3%	99.4%	99.3%
24	18	$0	$0	16.2%	22.8%	20.0%	19.8%	99.7%	99.7%	99.8%	99.7%
25	19	$0	$0	16.2%	22.8%	20.0%	19.7%	99.9%	99.9%	99.9%	99.9%
26	20	$0	$0	11.8%	16.7%	14.7%	14.5%	72.7%	73.2%	73.6%	73.2%
27	21	$0	$0	8.5%	12.1%	10.7%	10.5%	72.2%	72.3%	72.7%	72.4%
28	22	$0	$0	6.4%	9.0%	8.0%	7.8%	74.6%	74.3%	74.7%	74.5%
29	23	$0	$0	4.7%	6.5%	5.9%	5.7%	73.7%	72.9%	73.3%	73.3%
30	24	$0	$0	3.5%	4.8%	4.3%	4.2%	74.1%	73.1%	73.4%	73.5%
31	25	$0	$0	2.5%	3.4%	3.1%	3.0%	72.2%	70.9%	71.2%	71.3%
32	26	$0	$0	1.8%	2.4%	2.2%	2.2%	72.9%	71.4%	71.7%	71.9%
33	27	$0	$0	1.3%	1.6%	1.5%	1.5%	69.7%	68.0%	68.3%	68.6%
34	28	$0	$0	0.9%	1.1%	1.0%	1.0%	68.2%	66.6%	66.9%	67.1%
35	29	$0	$0	0.5%	0.6%	0.6%	0.6%	59.8%	58.3%	58.5%	58.8%
36	30	$0	$0	0.3%	0.3%	0.3%	0.3%	48.5%	47.3%	47.5%	47.7%

RAND *TR1260-A.2*

Figure A.3
Parameters Worksheet

	D	E	F	G–M
2	beta	0.90		"beta" is the discount rate
3	sigma	200		"sigma" > 0 is the standard deviation of the shock term (assumed to be normal with mean 0)
4	alpha	-30		"alpha" is the mode of the taste distribution (assumed to be extreme-value)
5	delta	15		"delta" > 0 is the scale of the taste distribution (assumed to be extreme-value)
6	alpha*ROTC	15		mode shift for ROTC
7	delta*ROTC	-5		scale shift for ROTC

RAND *TR1260-A.3*

```
' PDF and CDF for the standard Normal distribution

Function pnorm(ByRef X As Double) As Double
    pnorm = Application.WorksheetFunction.NormDist(X, 0, 1, True)
End Function

Function dnorm(ByRef X As Double) As Double
    dnorm = Application.WorksheetFunction.NormDist(X, 0, 1, False)
End Function
```

Figure A.4
Formulas in Columns A Through E of the SimpleDRM Worksheet

	A	B	C	D	E
1					
2				Academy CRR	ROTC CRR
3			ADSO	5	4
4			Fraction	=1260/4357	=1286/4357
5	Year of Service (YOS)	Bonus Pay	Separation Pay	Academy CRR	ROTC CRR
6	YOS	Wb	Ws		
7	1	1	0	=simulate1(Alpha,Delta,Sigma,Beta,D3,WbRange,WsRange)	=simulate1(Alpha+AlphaROTC,Delta+DeltaROTC,Sigma,Beta,E3,WbRange,WsRange)
8	2	0	0	=simulate1(Alpha,Delta,Sigma,Beta,D3,WbRange,WsRange)	=simulate1(Alpha+AlphaROTC,Delta+DeltaROTC,Sigma,Beta,E3,WbRange,WsRange)
9	3	0	0	=simulate1(Alpha,Delta,Sigma,Beta,D3,WbRange,WsRange)	=simulate1(Alpha+AlphaROTC,Delta+DeltaROTC,Sigma,Beta,E3,WbRange,WsRange)
10	4	0	0	=simulate1(Alpha,Delta,Sigma,Beta,D3,WbRange,WsRange)	=simulate1(Alpha+AlphaROTC,Delta+DeltaROTC,Sigma,Beta,E3,WbRange,WsRange)
11	5	0	0	=simulate1(Alpha,Delta,Sigma,Beta,D3,WbRange,WsRange)	=simulate1(Alpha+AlphaROTC,Delta+DeltaROTC,Sigma,Beta,E3,WbRange,WsRange)
12	6	0	0	=simulate1(Alpha,Delta,Sigma,Beta,D3,WbRange,WsRange)	=simulate1(Alpha+AlphaROTC,Delta+DeltaROTC,Sigma,Beta,E3,WbRange,WsRange)
13	7	0	0	=simulate1(Alpha,Delta,Sigma,Beta,D3,WbRange,WsRange)	=simulate1(Alpha+AlphaROTC,Delta+DeltaROTC,Sigma,Beta,E3,WbRange,WsRange)
14	8	0	0	=simulate1(Alpha,Delta,Sigma,Beta,D3,WbRange,WsRange)	=simulate1(Alpha+AlphaROTC,Delta+DeltaROTC,Sigma,Beta,E3,WbRange,WsRange)
15	9	0	0	=simulate1(Alpha,Delta,Sigma,Beta,D3,WbRange,WsRange)	=simulate1(Alpha+AlphaROTC,Delta+DeltaROTC,Sigma,Beta,E3,WbRange,WsRange)
16	10	0	0	=simulate1(Alpha,Delta,Sigma,Beta,D3,WbRange,WsRange)	=simulate1(Alpha+AlphaROTC,Delta+DeltaROTC,Sigma,Beta,E3,WbRange,WsRange)
17	11	0	0	=simulate1(Alpha,Delta,Sigma,Beta,D3,WbRange,WsRange)	=simulate1(Alpha+AlphaROTC,Delta+DeltaROTC,Sigma,Beta,E3,WbRange,WsRange)
18	12	0	0	=simulate1(Alpha,Delta,Sigma,Beta,D3,WbRange,WsRange)	=simulate1(Alpha+AlphaROTC,Delta+DeltaROTC,Sigma,Beta,E3,WbRange,WsRange)
19	13	0	0	=simulate1(Alpha,Delta,Sigma,Beta,D3,WbRange,WsRange)	=simulate1(Alpha+AlphaROTC,Delta+DeltaROTC,Sigma,Beta,E3,WbRange,WsRange)
20	14	0	0	=simulate1(Alpha,Delta,Sigma,Beta,D3,WbRange,WsRange)	=simulate1(Alpha+AlphaROTC,Delta+DeltaROTC,Sigma,Beta,E3,WbRange,WsRange)
21	15	0	0	=simulate1(Alpha,Delta,Sigma,Beta,D3,WbRange,WsRange)	=simulate1(Alpha+AlphaROTC,Delta+DeltaROTC,Sigma,Beta,E3,WbRange,WsRange)
22	16	0	0	=simulate1(Alpha,Delta,Sigma,Beta,D3,WbRange,WsRange)	=simulate1(Alpha+AlphaROTC,Delta+DeltaROTC,Sigma,Beta,E3,WbRange,WsRange)
23	17	0	0	=simulate1(Alpha,Delta,Sigma,Beta,D3,WbRange,WsRange)	=simulate1(Alpha+AlphaROTC,Delta+DeltaROTC,Sigma,Beta,E3,WbRange,WsRange)
24	18	0	0	=simulate1(Alpha,Delta,Sigma,Beta,D3,WbRange,WsRange)	=simulate1(Alpha+AlphaROTC,Delta+DeltaROTC,Sigma,Beta,E3,WbRange,WsRange)
25	19	0	0	=simulate1(Alpha,Delta,Sigma,Beta,D3,WbRange,WsRange)	=simulate1(Alpha+AlphaROTC,Delta+DeltaROTC,Sigma,Beta,E3,WbRange,WsRange)
26	20	0	0	=simulate1(Alpha,Delta,Sigma,Beta,D3,WbRange,WsRange)	=simulate1(Alpha+AlphaROTC,Delta+DeltaROTC,Sigma,Beta,E3,WbRange,WsRange)
27	21	0	0	=simulate1(Alpha,Delta,Sigma,Beta,D3,WbRange,WsRange)	=simulate1(Alpha+AlphaROTC,Delta+DeltaROTC,Sigma,Beta,E3,WbRange,WsRange)
28	22	0	0	=simulate1(Alpha,Delta,Sigma,Beta,D3,WbRange,WsRange)	=simulate1(Alpha+AlphaROTC,Delta+DeltaROTC,Sigma,Beta,E3,WbRange,WsRange)
29	23	0	0	=simulate1(Alpha,Delta,Sigma,Beta,D3,WbRange,WsRange)	=simulate1(Alpha+AlphaROTC,Delta+DeltaROTC,Sigma,Beta,E3,WbRange,WsRange)
30	24	0	0	=simulate1(Alpha,Delta,Sigma,Beta,D3,WbRange,WsRange)	=simulate1(Alpha+AlphaROTC,Delta+DeltaROTC,Sigma,Beta,E3,WbRange,WsRange)
31	25	0	0	=simulate1(Alpha,Delta,Sigma,Beta,D3,WbRange,WsRange)	=simulate1(Alpha+AlphaROTC,Delta+DeltaROTC,Sigma,Beta,E3,WbRange,WsRange)
32	26	0	0	=simulate1(Alpha,Delta,Sigma,Beta,D3,WbRange,WsRange)	=simulate1(Alpha+AlphaROTC,Delta+DeltaROTC,Sigma,Beta,E3,WbRange,WsRange)
33	27	0	0	=simulate1(Alpha,Delta,Sigma,Beta,D3,WbRange,WsRange)	=simulate1(Alpha+AlphaROTC,Delta+DeltaROTC,Sigma,Beta,E3,WbRange,WsRange)
34	28	0	0	=simulate1(Alpha,Delta,Sigma,Beta,D3,WbRange,WsRange)	=simulate1(Alpha+AlphaROTC,Delta+DeltaROTC,Sigma,Beta,E3,WbRange,WsRange)
35	29	0	0	=simulate1(Alpha,Delta,Sigma,Beta,D3,WbRange,WsRange)	=simulate1(Alpha+AlphaROTC,Delta+DeltaROTC,Sigma,Beta,E3,WbRange,WsRange)
36	30	0	0	=simulate1(Alpha,Delta,Sigma,Beta,D3,WbRange,WsRange)	=simulate1(Alpha+AlphaROTC,Delta+DeltaROTC,Sigma,Beta,E3,WbRange,WsRange)

RAND *TR1260-A.4*

Figure A.5
Formulas in Columns F Through K of the SimpleDRM Worksheet

	F	G	H	I	J	K
1						
2	Non-Scholarship ROTC CRR					
3	3					
4	=1811/4357					
5	Non-Scholarship ROTC CRR	CRR	Academy Y2Y	ROTC Y2Y	Non-Scholarship ROTC Y2Y	Y2Y
6						
7	=simulate1(Alpha+AlphaROTC,Delta+DeltaROTC,Sigma,Beta,F3,WbRange,WsRange)	=SUMPRODUCT(D$4:F$4,D7:F7)	=D7	=E7	=F7	=1
8	=simulate1(Alpha+AlphaROTC,Delta+DeltaROTC,Sigma,Beta,F3,WbRange,WsRange)	=SUMPRODUCT(D$4:F$4,D8:F8)	=D8/D7	=E8/E7	=F8/F7	=G8/G7
9	=simulate1(Alpha+AlphaROTC,Delta+DeltaROTC,Sigma,Beta,F3,WbRange,WsRange)	=SUMPRODUCT(D$4:F$4,D9:F9)	=D9/D8	=E9/E8	=F9/F8	=G9/G8
10	=simulate1(Alpha+AlphaROTC,Delta+DeltaROTC,Sigma,Beta,F3,WbRange,WsRange)	=SUMPRODUCT(D$4:F$4,D10:F10)	=D10/D9	=E10/E9	=F10/F9	=G10/G9
11	=simulate1(Alpha+AlphaROTC,Delta+DeltaROTC,Sigma,Beta,F3,WbRange,WsRange)	=SUMPRODUCT(D$4:F$4,D11:F11)	=D11/D10	=E11/E10	=F11/F10	=G11/G10
12	=simulate1(Alpha+AlphaROTC,Delta+DeltaROTC,Sigma,Beta,F3,WbRange,WsRange)	=SUMPRODUCT(D$4:F$4,D12:F12)	=D12/D11	=E12/E11	=F12/F11	=G12/G11
13	=simulate1(Alpha+AlphaROTC,Delta+DeltaROTC,Sigma,Beta,F3,WbRange,WsRange)	=SUMPRODUCT(D$4:F$4,D13:F13)	=D13/D12	=E13/E12	=F13/F12	=G13/G12
14	=simulate1(Alpha+AlphaROTC,Delta+DeltaROTC,Sigma,Beta,F3,WbRange,WsRange)	=SUMPRODUCT(D$4:F$4,D14:F14)	=D14/D13	=E14/E13	=F14/F13	=G14/G13
15	=simulate1(Alpha+AlphaROTC,Delta+DeltaROTC,Sigma,Beta,F3,WbRange,WsRange)	=SUMPRODUCT(D$4:F$4,D15:F15)	=D15/D14	=E15/E14	=F15/F14	=G15/G14
16	=simulate1(Alpha+AlphaROTC,Delta+DeltaROTC,Sigma,Beta,F3,WbRange,WsRange)	=SUMPRODUCT(D$4:F$4,D16:F16)	=D16/D15	=E16/E15	=F16/F15	=G16/G15
17	=simulate1(Alpha+AlphaROTC,Delta+DeltaROTC,Sigma,Beta,F3,WbRange,WsRange)	=SUMPRODUCT(D$4:F$4,D17:F17)	=D17/D16	=E17/E16	=F17/F16	=G17/G16
18	=simulate1(Alpha+AlphaROTC,Delta+DeltaROTC,Sigma,Beta,F3,WbRange,WsRange)	=SUMPRODUCT(D$4:F$4,D18:F18)	=D18/D17	=E18/E17	=F18/F17	=G18/G17
19	=simulate1(Alpha+AlphaROTC,Delta+DeltaROTC,Sigma,Beta,F3,WbRange,WsRange)	=SUMPRODUCT(D$4:F$4,D19:F19)	=D19/D18	=E19/E18	=F19/F18	=G19/G18
20	=simulate1(Alpha+AlphaROTC,Delta+DeltaROTC,Sigma,Beta,F3,WbRange,WsRange)	=SUMPRODUCT(D$4:F$4,D20:F20)	=D20/D19	=E20/E19	=F20/F19	=G20/G19
21	=simulate1(Alpha+AlphaROTC,Delta+DeltaROTC,Sigma,Beta,F3,WbRange,WsRange)	=SUMPRODUCT(D$4:F$4,D21:F21)	=D21/D20	=E21/E20	=F21/F20	=G21/G20
22	=simulate1(Alpha+AlphaROTC,Delta+DeltaROTC,Sigma,Beta,F3,WbRange,WsRange)	=SUMPRODUCT(D$4:F$4,D22:F22)	=D22/D21	=E22/E21	=F22/F21	=G22/G21
23	=simulate1(Alpha+AlphaROTC,Delta+DeltaROTC,Sigma,Beta,F3,WbRange,WsRange)	=SUMPRODUCT(D$4:F$4,D23:F23)	=D23/D22	=E23/E22	=F23/F22	=G23/G22
24	=simulate1(Alpha+AlphaROTC,Delta+DeltaROTC,Sigma,Beta,F3,WbRange,WsRange)	=SUMPRODUCT(D$4:F$4,D24:F24)	=D24/D23	=E24/E23	=F24/F23	=G24/G23
25	=simulate1(Alpha+AlphaROTC,Delta+DeltaROTC,Sigma,Beta,F3,WbRange,WsRange)	=SUMPRODUCT(D$4:F$4,D25:F25)	=D25/D24	=E25/E24	=F25/F24	=G25/G24
26	=simulate1(Alpha+AlphaROTC,Delta+DeltaROTC,Sigma,Beta,F3,WbRange,WsRange)	=SUMPRODUCT(D$4:F$4,D26:F26)	=D26/D25	=E26/E25	=F26/F25	=G26/G25
27	=simulate1(Alpha+AlphaROTC,Delta+DeltaROTC,Sigma,Beta,F3,WbRange,WsRange)	=SUMPRODUCT(D$4:F$4,D27:F27)	=D27/D26	=E27/E26	=F27/F26	=G27/G26
28	=simulate1(Alpha+AlphaROTC,Delta+DeltaROTC,Sigma,Beta,F3,WbRange,WsRange)	=SUMPRODUCT(D$4:F$4,D28:F28)	=D28/D27	=E28/E27	=F28/F27	=G28/G27
29	=simulate1(Alpha+AlphaROTC,Delta+DeltaROTC,Sigma,Beta,F3,WbRange,WsRange)	=SUMPRODUCT(D$4:F$4,D29:F29)	=D29/D28	=E29/E28	=F29/F28	=G29/G28
30	=simulate1(Alpha+AlphaROTC,Delta+DeltaROTC,Sigma,Beta,F3,WbRange,WsRange)	=SUMPRODUCT(D$4:F$4,D30:F30)	=D30/D29	=E30/E29	=F30/F29	=G30/G29
31	=simulate1(Alpha+AlphaROTC,Delta+DeltaROTC,Sigma,Beta,F3,WbRange,WsRange)	=SUMPRODUCT(D$4:F$4,D31:F31)	=D31/D30	=E31/E30	=F31/F30	=G31/G30
32	=simulate1(Alpha+AlphaROTC,Delta+DeltaROTC,Sigma,Beta,F3,WbRange,WsRange)	=SUMPRODUCT(D$4:F$4,D32:F32)	=D32/D31	=E32/E31	=F32/F31	=G32/G31
33	=simulate1(Alpha+AlphaROTC,Delta+DeltaROTC,Sigma,Beta,F3,WbRange,WsRange)	=SUMPRODUCT(D$4:F$4,D33:F33)	=D33/D32	=E33/E32	=F33/F32	=G33/G32
34	=simulate1(Alpha+AlphaROTC,Delta+DeltaROTC,Sigma,Beta,F3,WbRange,WsRange)	=SUMPRODUCT(D$4:F$4,D34:F34)	=D34/D33	=E34/E33	=F34/F33	=G34/G33
35	=simulate1(Alpha+AlphaROTC,Delta+DeltaROTC,Sigma,Beta,F3,WbRange,WsRange)	=SUMPRODUCT(D$4:F$4,D35:F35)	=D35/D34	=E35/E34	=F35/F34	=G35/G34
36	=simulate1(Alpha+AlphaROTC,Delta+DeltaROTC,Sigma,Beta,F3,WbRange,WsRange)	=SUMPRODUCT(D$4:F$4,D36:F36)	=D36/D35	=E36/E35	=F36/F35	=G36/G35

RAND *TR1260-A.5*

```
' Quantile and density functions for the extreme value distribution

Function qev(ByRef X As Double, ByRef Alpha As Double, _
  ByRef Delta As Double) As Double
  qev = Alpha - Delta * Log(Log(1 / X))
End Function
Function dev(ByRef X As Double, ByRef Alpha As Double, _
  ByRef Delta As Double) As Double
  dev = Exp(-Exp((Alpha - X) / Delta) + (Alpha - X) / Delta) / Delta
End Function

Function Wc(ByRef YOS As Integer) As Double
 ' Initialize civilian wage array
  Static NotFirstTime As Boolean
  If NotFirstTime = False Then
        Static WcArray As Variant
        WcArray = Array(67.105792, 67.105792, 67.105792, _
              67.105792, 67.105792, 65.365508, 72.423728, _
              81.6933, 86.394464, 91.64142, 97.69786, _
              105.905748, 106.705144, 114.378264, 111.475468, _
              116.178816, 114.824684, 115.87056, 118.419652, _
              117.231816, 120.493152, 119.887092, 119.815696, _
              120.620292, 119.350868, 121.093232, 123.364436, _
              117.889408, 118.280084, 118.280084, 118.280084, _
              118.280084, 118.280084, 118.280084, 118.280084)
        NotFirstTime = True
  End If
  Wc = WcArray(YOS)
End Function

Function Wm(ByRef YOS As Integer) As Double
  ' Initialize military wage array
  Static NotFirstTime As Boolean
  If NotFirstTime = False Then
        Static WmArray As Variant
        WmArray = Array(55.353, 65.389, 73.815, 81.184, _
              81.184, 85.256, 85.256, 88.771, 88.771, 96.074, _
              96.074, 101.773, 101.773, 105.804, 105.804, _
              111.965, 111.965, 117.385, 117.385, 121.351, _
              121.351, 128.671, 128.671, 133.202, 133.202, _
              139.669, 139.669, 144.23, 144.23, 148.37, 0#)
        NotFirstTime = True
  End If
  Wm = WmArray(YOS)
End Function
```

```
Function Wb(ByRef YOS As Integer) As Double
   'Bonus Pay
   Dim range1 As Range
   Dim range2 As Range
   Set range1 = Worksheets("SimpleDRM").Range("YOSRange")
   Set range2 = Worksheets("SimpleDRM").Range("WbRange")
   Wb = Application.WorksheetFunction.Lookup(YOS, range1, range2) / 1000
End Function

Function Ws(ByRef YOS As Integer) As Double
   'Separation Pay
   Dim range1 As Range
   Dim range2 As Range
   Set range1 = Worksheets("SimpleDRM").Range("YOSRange")
   Set range2 = Worksheets("SimpleDRM").Range("WsRange")
   Ws = Application.WorksheetFunction.Lookup(YOS, range1, range2) / 1000
End Function

Function RetirementPay(ByRef YOS As Integer, ByRef Beta As Double) As Double
   If YOS < 20 Then
         RetirementPay = 0#
   Else
         'Infinite series approximation to DPV of retirement pay
         RetirementPay = (1# / (1# - Beta)) * _
               (((1# / 2#) + (YOS - 21#) * (1# / 40#)) * _
               ((Wm(YOS - 1) + Wm(YOS - 2) + Wm(YOS - 3)) / 3#))
   End If
End Function

Function SeparationPay(ByRef YOS As Integer) As Double
   SeparationPay = Ws(YOS)
End Function
```

Basic Dynamic Retention Model

The basic DRM implements a version of the model in which the only choice confronting an officer from year to year is whether to stay for an additional year or to leave. The VBA code, given a set Gamma and YOS, calculates the nonstochastic value of staying for one more year, **Vs** and the value of leaving, **Vl**; subtracts the value of leaving from the value of staying; and calculates the probability of staying given size of the difference and the standard deviation of the environmental disturbance **Sigma**.

The nonstochastic value of staying in a particular YOS, **Vs(YOS, Gamma, Sigma, Beta)**, is the sum of the individual's taste **Gamma**, the military wage corresponding to the YOS, **Wm(YOS)**, the bonus pay for that YOS, **Wb(YOS)**, and the discounted expected value of the maximum of staying or leaving in the following YOS, **Beta * EV(YOS + 1, Gamma, Sigma,**

Beta). The value of leaving in a particular YOS, **Vl(YOS, Beta)**, is the discounted present value of the stream of payments **Wc** from a civilian career starting in a given year of service, plus the discounted present value of the stream of military retirement benefits for serving until a given year of service, **RetirementPay(YOS, Beta)**, plus the separation pay received for separating in a given year of service, **Ws(YOS)**.

The **Simulatel** function calculates the cumulative probability that an officer stays for each YOS from the year when the officer has completed his or her active-duty service requirement (which typically depends on the source of accession). The **Simulatel** function also integrates out the unobserved heterogeneity, **Gamma**, given that **Gamma** is extreme-value distributed with mode parameter **Alpha** and scale parameter **Delta**. **Simulatel** returns a vector of cumulative stay probabilities given the parameters of the taste distribution, the disturbance distribution, the personal discount rate, and the duration of the initial active-duty service obligation.

Program Listing for the Basic Dynamic Retention Model

```
' Value of staying one more year, net of the environmental disturbance

Function Vs(ByRef YOS As Integer, ByRef Gamma As Double, _
  ByRef Sigma As Double, ByRef Beta As Double) As Double
  If YOS = T Then
        Vs = 0#
  Else
        Vs = Gamma + Wm(YOS) + Wb(YOS) + _
              Beta * EV(YOS + 1, Gamma, Sigma, Beta)
  End If
End Function

' Value of leaving

Function Vl(ByRef YOS As Integer, ByRef Beta As Double) As Double
  If YOS = T Then
        Vl = Wc(YOS) + RetirementPay(YOS, Beta)
  Else
        Vl = 0#
        Dim s As Integer
        For s = YOS To T
              Vl = Vl + (Beta ^ (s - YOS)) * Wc(s)
        Next s
        Vl = Vl + RetirementPay(YOS, Beta) + SeparationPay(YOS)
  End If
End Function

' Expected value of the maximum of staying or leaving,
' taking into account the future environmental disturbance

Function EV(ByRef YOS As Integer, ByRef Gamma As Double, _
        ByRef Sigma As Double, ByRef Beta As Double) As Double
```

```
    Dim Vs1 As Double
    Dim Vl1 As Double
    Dim a As Double
    Dim F As Double

    Vs1 = Vs(YOS, Gamma, Sigma, Beta)
    Vl1 = Vl(YOS, Beta)
    a = (Vs1 - Vl1) / Sigma
    F = pnorm(a)
    EV = F * Vs1 + (1 - F) * Vl1 + Sigma * dnorm(a)
End Function

' Year-to-year probability of staying

Function PrStay(ByRef Gamma As Double, ByRef Sigma As Double, _
            ByRef Beta As Double, ByRef YOS As Integer) As Double
    PrStay = pnorm((Vs(YOS, Gamma, Sigma, Beta) - Vl(YOS, Beta)) / Sigma)
End Function

' Note that Simulate1 is an Array function
' Simulate1 calculates the cumulative probability of staying to a certain YOS,
' integrating out unobserved heterogeneity

Function Simulate1(ByRef Alpha As Double, ByRef Delta As Double, _
    ByRef Sigma As Double, ByRef Beta As Double, _
    ByRef YOSMin As Integer, ByRef WbRange As Range, _
    ByRef WsRange As Range) As Variant
    Dim gammaMin As Double
    Dim gammaMax As Double
    Dim gammaStep As Double
    Dim p As Double
    Dim p1(30, 1) As Double
    Dim p2(30, 1) As Double
    Dim normalizingConstant As Double

    gammaMin = qev(0.01, Alpha, Delta)
    gammaMax = qev(0.99, Alpha, Delta)
    gammaStep = (gammaMax - gammaMin) / 34
    normalizingConstant = 0

    Dim Gamma As Double
    Dim MyYOS As Integer

    For Gamma = gammaMin To gammaMax Step gammaStep
            p = 1#
            normalizingConstant = normalizingConstant + dev(Gamma, Alpha, Delta)
            For MyYOS = YOSMin + 1 To 30
```

```
            p = p * PrStay(Gamma, Sigma, Beta, MyYOS)
            p1(MyYOS, 1) = p1(MyYOS, 1) + p * dev(Gamma, Alpha, Delta)
        Next MyYOS
    Next Gamma

    For MyYOS = 1 To YOSMin
        p2(MyYOS, 1) = 1
    Next MyYOS

    For MyYOS = YOSMin + 1 To 30
        p2(MyYOS, 1) = p1(MyYOS, 1) / normalizingConstant
    Next MyYOS

    Simulate1 = p2

End Function
```

The Air Force Aviator Continuation Pay Program

The ACP spreadsheet DRM extends the basic model to cover the case in which an initially eligible officer can choose between serving from year to year and signing up for the ACP program, a five-year obligation in exchange for a bonus of $25,000 per year for each year he or she is obligated. Once the five-year obligation is completed, the officer serves from year to year. So, in addition to calculating the nonstochastic value of staying one more year, **Vs**, and the value of leaving, **Vl**, the program also calculates the value of staying for five years, **Vs5**, for those officers who are initially eligible. The program calculates **Vs** and **Vs5** for a particular **Gamma** and subtracts **Vl** from the maximum of the two to compute the probability of an initially eligible officer staying. If **Vs5** is greater than **Vs** for a particular **Gamma**, then the program sets the **TookACP** flag to **True** and starts the **CountdownACP** timer, which counts down the years of service until the officer is eligible to make another stay-or-leave decision. Once the timer has reached zero, the program calculates the year-to-year probability of staying. If **Vs5** is not greater than **Vs**, then the program simply computes the year-to-year probability of staying, as in the basic DRM.

Program Listing for the Air Force Aviator Continuation Pay Program

```
Function Vs(ByRef YOS As Integer, ByRef Gamma As Double, _
  ByRef Sigma As Double, ByRef Beta As Double) As Double
  If YOS = t Then
        Vs = 0#
  Else
        Vs = Gamma + Wm(YOS) + Wb(YOS) + _
        Beta * EV(YOS + 1, Gamma, Sigma, Beta)
  End If
End Function
```

```
Function Vs5(ByRef YOS As Integer, ByRef Gamma As Double, _
   ByRef Sigma As Double, ByRef Beta As Double) As Double
   Dim temp As Double
   Dim s As Integer
   If YOS = t Then
         Vs5 = 0#
   Else
         temp = 0
         For s = 0 To 4
               temp = temp + _
                     Beta ^ s * (Gamma + Wm(YOS + s) + Wb(YOS + s) + 25)
         Next s
         Vs5 = temp + (Beta ^ 5) * EV(YOS + 5, Gamma, Sigma, Beta)
   End If
End Function

Function Vl(ByRef YOS As Integer, ByRef Beta As Double) As Double
   If YOS = t Then
         Vl = Wc(YOS) + RetirementPay(YOS, Beta)
   Else
         Vl = 0#
         Dim s As Integer
         For s = YOS To t
               Vl = Vl + (Beta ^ (s - YOS)) * Wc(s)
         Next s
         Vl = Vl + RetirementPay(YOS, Beta) + SeparationPay(YOS)
   End If
End Function

Function EV(YOS As Integer, Gamma As Double, Sigma As Double, _
   Beta As Double) As Double
   Dim Vs1 As Double
   Dim Vl1 As Double
   Dim a As Double
   Dim F As Double

   Vs1 = Vs(YOS, Gamma, Sigma, Beta)
   Vl1 = Vl(YOS, Beta)
   a = (Vs1 - Vl1) / Sigma
   F = pnorm(a)
   EV = F * Vs1 + (1 - F) * Vl1 + Sigma * dnorm(a)
End Function

Function PrStay(ByRef Gamma As Double, ByRef Sigma As Double, _
   ByRef Beta As Double, ByRef YOS As Integer) As Double
   PrStay = pnorm((Vs(YOS, Gamma, Sigma, Beta) - Vl(YOS, Beta)) / Sigma)
```

```
End Function

Function PrStay5(ByRef Gamma As Double, ByRef Sigma As Double, _
  ByRef Beta As Double, ByRef YOS As Integer) As Double
  PrStay5 = pnorm((Vs5(YOS, Gamma, Sigma, Beta) - Vl(YOS, Beta)) / Sigma)
End Function

' Note that Simulate is an Array function
Function Simulate1(ByRef Alpha As Double, ByRef Delta As Double, _
  ByRef Sigma As Double, ByRef Beta As Double, _
  ByRef YOSMin As Integer, ByRef WbRange As Range, _
  ByRef WsRange As Range) As Variant

  Dim gammaMin As Double
  Dim gammaMax As Double
  Dim gammaStep As Double
  Dim p As Double
  Dim p1(30, 1) As Double
  Dim p2(30, 1) As Double
  Dim normalizingConstant As Double

  gammaMin = qev(0.01, Alpha, Delta)
  gammaMax = qev(0.99, Alpha, Delta)
  gammaStep = (gammaMax - gammaMin) / 34
  normalizingConstant = 0

  Dim Gamma As Double
  Dim MyYOS As Integer

  Dim TookACP As Boolean
  Dim CountdownACP As Integer

  For Gamma = gammaMin To gammaMax Step gammaStep
        p = 1#
        TookACP = False
        CountdownACP = 0
        normalizingConstant = normalizingConstant + _
              dev(Gamma, Alpha, Delta)

        For MyYOS = YOSMin + 1 To 30
              If CountdownACP > 0 Then
                    CountdownACP = CountdownACP - 1
                    'p = p * 1#
              ElseIf TookACP = True Then
                    p = p * PrStay(Gamma, Sigma, Beta, MyYOS)
              ElseIf MyYOS = YOSMin + 1 Then
                    If (Vs5(MyYOS, Gamma, Sigma, Beta) > _
```

```
                            Vs(MyYOS, Gamma, Sigma, Beta)) Then
                                TookACP = True
                                CountdownACP = 4
                                p = PrStay5(Gamma, Sigma, Beta, MyYOS)
                        Else
                                p = PrStay(Gamma, Sigma, Beta, MyYOS)
                        End If
                Else
                        p = p * PrStay(Gamma, Sigma, Beta, MyYOS)
                End If
                p1(MyYOS, 1) = p1(MyYOS, 1) + p * dev(Gamma, Alpha, Delta)
        Next MyYOS
    Next Gamma

    For MyYOS = 1 To YOSMin
        p2(MyYOS, 1) = 1
    Next MyYOS

    For MyYOS = YOSMin + 1 To 30
        p2(MyYOS, 1) = p1(MyYOS, 1) / normalizingConstant
    Next MyYOS

    Simulate1 = p2

End Function
```

The Army Graduate School for Service Program

The Army GRADSO program spreadsheet DRM extends the basic model to examine the effect of offering the option to attend graduate school to those officers who agree to extend their commissioning active-duty service obligation by three years. This extension is called GRADSO. Once the three-year GRADSO is completed, the officer has the option of either attending graduate school (which is paid for by the Army) and incurring an additional service requirement of 3:1 for every year in graduate school, serving from year to year, or leaving. So, in addition to calculating the nonstochastic value of staying one more year, **Vs1**, and the value of leaving, **Vl**, the program also calculates the value of staying, extending the initial commissioning active-duty service requirement by three years, and getting the option to attend graduate school, **Vg**. (In this version of the model, we assume the duration of graduate school to always be two years, which means that the decision to attend graduate school results in a payback obligation of six years, resulting in a total of eight additional years of service.) For those officers who complete GRADSO, the program calculates the value of taking the graduate school option, **Vd**. The program calculates **Vs1** and **Vg** for a particular **Gamma** and subtracts **Vl** from the maximum of the two to compute the probability of an officer staying. If **Vg** is greater than **Vs1** for a particular **Gamma**, then the program sets the **TookGRADSO** flag to **True** and starts the **CountdownGRADSO** timer, which counts down the years of service until the officer is eligible to decide to attend graduate school, serve from year to year, or leave. Once the **CountdownGRADSO** timer has reached zero, the program calculates the probability of staying by calculating **Vs1** and **Vd** for the officer's **Gamma** and subtracts **Vl** from the maximum of the two to compute the probability of an officer staying. If **Vd** is greater than **Vs1**, then the program sets the **TookDegree** flag to **True** and starts the **CountdownDegree** timer, which counts down the years of service until the officer is eligible to decide to serve from year to year or leave. If **Vd** is not greater than **Vs**, then the program simply computes the year-to-year probability of staying, as in the basic DRM.

Program Listing for the Army Graduate School for Service Program

```
Function Vs1(ByRef YOS As Integer, ByRef Gamma As Double, _
   ByRef Sigma As Double, ByRef Beta As Double) As Double
   If YOS = t Then
         Vs1 = 0#
   Else
         Vs1 = Gamma + Wm(YOS) + Wb(YOS) + _
         Beta * EmaxVs1Vl(YOS + 1, Gamma, Sigma, Beta)
   End If
End Function

Function Vg(ByRef YOS As Integer, ByRef Gamma As Double, _
         ByRef Sigma As Double, ByRef Beta As Double) As Double
   Dim temp As Double
   Dim s As Integer
   temp = 0
   For s = 0 To 2
         temp = temp + Beta ^ s * (Gamma + Wm(YOS + s) + Wb(YOS + s))
   Next s
   Vg = temp + (Beta ^ 3) * EmaxVs1VdVl(YOS + 5, Gamma, Sigma, Beta)
End Function

Function Vd(ByRef YOS As Integer, ByRef Gamma As Double, _
   ByRef Sigma As Double, ByRef Beta As Double) As Double
   'GammaD is the DPV of graduate school and lifecycle increase in earnings
   Dim GammaD As Double
   GammaD = 100

   Dim temp As Double
   Dim s As Integer

   temp = 0
   For s = 0 To 7
         temp = temp + Beta ^ s * (Gamma + Wm(YOS + s) + Wb(YOS + s))
   Next s
   Vd = GammaD + temp + (Beta ^ 8) * EmaxVs1Vl(YOS + 8, Gamma, Sigma, Beta)
End Function
```

```
Function Vl(ByRef YOS As Integer, ByRef Beta As Double) As Double
   If YOS = t Then
         Vl = Wc(YOS) + RetirementPay(YOS, Beta)
   Else
         Vl = 0#
         Dim s As Integer
         For s = YOS To t
               Vl = Vl + (Beta ^ (s - YOS)) * Wc(s)
         Next s
         Vl = Vl + RetirementPay(YOS, Beta) + SeparationPay(YOS)
   End If
End Function

Function EmaxVs1Vl(YOS As Integer, Gamma As Double, _
   Sigma As Double, Beta As Double) As Double
   Dim Vs1x As Double
   Dim Vlx As Double
   Dim a As Double
   Dim F As Double

   Vs1x = Vs1(YOS, Gamma, Sigma, Beta)
   Vlx = Vl(YOS, Beta)
   a = (Vs1x - Vlx) / Sigma
   F = pnorm(a)
   EmaxVs1Vl = F * Vs1x + (1 - F) * Vlx + Sigma * dnorm(a)
End Function

Function EmaxVs1VdVl(YOS As Integer, Gamma As Double, _
   Sigma As Double, Beta As Double) As Double
   Dim Vs1x As Double
   Dim Vdx As Double
   Dim Vlx As Double
   Dim Vmax As Double
   Dim a As Double
   Dim F As Double

   Vs1x = Vs1(YOS, Gamma, Sigma, Beta)
   Vdx = Vd(YOS, Gamma, Sigma, Beta)
   Vlx = Vl(YOS, Beta)
   Vmax = Application.WorksheetFunction.Max(Vs1x, Vdx)
   a = (Vmax - Vlx) / Sigma
   F = pnorm(a)
   EmaxVs1VdVl = F * Vmax + (1 - F) * Vlx + Sigma * dnorm(a)
End Function

Function PrStay1(ByRef Gamma As Double, ByRef Sigma As Double, _
```

```
    ByRef Beta As Double, ByRef YOS As Integer) As Double
    PrStay1 = pnorm((Vs1(YOS, Gamma, Sigma, Beta) - Vl(YOS, Beta)) / Sigma)
End Function

Function PrStayG(ByRef Gamma As Double, ByRef Sigma As Double, _
    ByRef Beta As Double, ByRef YOS As Integer) As Double
    PrStayG = pnorm((Vg(YOS, Gamma, Sigma, Beta) - Vl(YOS, Beta)) / Sigma)
End Function

Function PrStayD(ByRef Gamma As Double, ByRef Sigma As Double, _
    ByRef Beta As Double, ByRef YOS As Integer) As Double
    PrStayD = pnorm((Vd(YOS, Gamma, Sigma, Beta) - Vl(YOS, Beta)) / Sigma)
End Function

'Note that Simulate is an Array function
Function Simulate1(ByRef Alpha As Double, ByRef Delta As Double, _
    ByRef Sigma As Double, ByRef Beta As Double, _
    ByRef YOSMin As Integer, ByRef WbRange As Range, _
    ByRef WsRange As Range) As Variant

    Dim gammaMin As Double
    Dim gammaMax As Double
    Dim gammaStep As Double
    Dim p As Double
    Dim p1(30, 1) As Double
    Dim p2(30, 1) As Double
    Dim normalizingConstant As Double

    gammaMin = qev(0.01, Alpha, Delta)
    gammaMax = qev(0.99, Alpha, Delta)
    gammaStep = (gammaMax - gammaMin) / 34
    normalizingConstant = 0

    Dim Gamma As Double
    Dim MyYOS As Integer

    Dim TookGRADSO As Boolean
    Dim CountdownGRADSO As Integer

    Dim TookDegree As Boolean
    Dim CountdownDegree As Integer

For Gamma = gammaMin To gammaMax Step gammaStep
    p = 1#
    TookGRADSO = False
    CountdownGRADSO = 0
    TookDegree = False
```

```
CountdownDegree = 0
normalizingConstant = normalizingConstant + _
        dev(Gamma, Alpha, Delta)

For MyYOS = YOSMin + 1 To 30
                If CountdownGRADSO > 0 Then
                        CountdownGRADSO = CountdownGRADSO - 1
                        'p = p * 1#
                ElseIf CountdownDegree > 0 Then
                        CountdownDegree = CountdownDegree - 1
                        'p = p * 1#
                ElseIf MyYOS = YOSMin + 4 Then
                        If TookGRADSO = True Then
                                If (Vd(MyYOS, Gamma, Sigma, Beta) > _
                                        Vs1(MyYOS, Gamma, Sigma, Beta)) Then
                                        TookDegree = True
                                        CountdownDegree = 7
                                        p = p * PrStayD(Gamma, Sigma, Beta, MyYOS)
                                Else
                                        p = p * PrStay1(Gamma, Sigma, Beta, MyYOS)
                                End If
                        Else
                                p = p * PrStay1(Gamma, Sigma, Beta, MyYOS)
                        End If
                ElseIf MyYOS = YOSMin + 1 Then
                        If (Vg(MyYOS, Gamma, Sigma, Beta) > _
                                Vs1(MyYOS, Gamma, Sigma, Beta)) Then
                                TookGRADSO = True
                                CountdownGRADSO = 2
                                p = PrStayG(Gamma, Sigma, Beta, MyYOS)
                        Else
                                p = PrStay1(Gamma, Sigma, Beta, MyYOS)
                        End If
                Else
                        p = p * PrStay1(Gamma, Sigma, Beta, MyYOS)
                End If
                p1(MyYOS, 1) = p1(MyYOS, 1) + p * dev(Gamma, Alpha, Delta)
        Next MyYOS
Next Gamma

For MyYOS = 1 To YOSMin
        p2(MyYOS, 1) = 1
Next MyYOS

For MyYOS = YOSMin + 1 To 30
        p2(MyYOS, 1) = p1(MyYOS, 1) / normalizingConstant
```

```
    Next MyYOS
    Simulate1 = p2
End Function
```

Bibliography

Asch, Beth J., James Hosek, and Craig Martin, *A Look at Cash Compensation for Active-Duty Military Personnel*, Santa Monica, Calif.: RAND Corporation, MR-1492-OSD, 2002. As of July 8, 2013: http://www.rand.org/pubs/monograph_reports/MR1492.html

Asch, Beth J., James Hosek, Michael G. Mattock, and Christina Panis, *Assessing Compensation Reform: Research in Support of the 10th Quadrennial Review of Military Compensation*, Santa Monica, Calif.: RAND Corporation, MG-764-OSD, 2008. As of August 29, 2012: http://www.rand.org/pubs/monographs/MG764.html

Asch, Beth J., and John T. Warner, *A Theory of Military Compensation and Personnel Policy*, Santa Monica, Calif.: RAND Corporation, MR-439-OSD, 1994. As of August 29, 2012: http://www.rand.org/pubs/monograph_reports/MR439.html

Butler, J. S., and Robert Moffitt, "A Computationally Efficient Quadrature Procedure for the One-Factor Multinomial Probit Model," *Econometrica*, Vol. 50, No. 3, May 1982, pp. 761–764.

Campion, MAJ Greg, U.S. Army Office of Economic and Manpower Analysis, *Post 9/11 "NEW" GI Bill and Career Incentive Programs: Class of 2010*, briefing slides, 2009.

Clark, Charles E., "The Greatest of a Finite Set of Random Variables," *Operations Research*, Vol. 9, No. 2, March–April 1961, pp. 145–162.

Daula, Thomas, and Robert Moffitt, "Estimating Dynamic Models of Quit Behavior: The Case of Military Reenlistment," *Journal of Labor Economics*, Vol. 13, No. 3, July 1995, pp. 499–523.

DeBacker, Jason, Bradley Heim, Vasia Panousi, and Ivan Vidangos, "Rising Inequality: Transitory or Permanent? New Evidence of a Panel of U.S. Tax Returns 1987–2006," Bloomington, Ind.: Indiana University–Bloomington School of Public and Environmental Affairs Research Paper Series 2011-01-01, January 2, 2012.

Department of the Army, Officer Active Duty Service Obligations, Washington, D.C., Army Regulation 350-100, August 8, 2007, revised August 10, 2009. As of August 29, 2012: http://www.apd.army.mil/pdffiles/r350_100.pdf

DoD—*See* U.S. Department of Defense.

Fricker, Ronald D., *The Effects of Perstempo on Officer Retention in the U.S. Military*, Santa Monica, Calif.: RAND Corporation, MR-1556-OSD, 2002. As of July 8, 2013: http://www.rand.org/pubs/monograph_reports/MR1556.html

Gotz, Glenn A., and John McCall, *A Dynamic Retention Model for Air Force Officers: Theory and Estimates*, Santa Monica, Calif.: RAND Corporation, R-3028-AF, 1984. As of August 29, 2012: http://www.rand.org/pubs/reports/R3028.html

Hosek, James, and Paco Martorell, *How Have Deployments During the War on Terrorism Affected Reenlistment?* Santa Monica, Calif.: RAND Corporation, MG-873-OSD, 2009. As of July 8, 2013: http://www.rand.org/pubs/monographs/MG873.html

Hosek, James, Michael G. Mattock, C. Christine Fair, Jennifer Kavanagh, Jennifer Sharp, and Mark E. Totten, *Attracting the Best: How the Military Competes for Information Technology Personnel*, Santa Monica, Calif.: RAND Corporation, MG-108-OSD, 2004. As of August 29, 2012: http://www.rand.org/pubs/monographs/MG108.html

Hosek, James, and Mark E. Totten, *Serving Away from Home: How Deployments Influence Reenlistment*, Santa Monica, Calif.: RAND Corporation, MR-1594-OSD, 2002. As of July 8, 2013: http://www.rand.org/pubs/monograph_reports/MR1594.html

Lyle, David S., "Using Military Deployments and Job Assignments to Estimate the Effect of Parental Absences and Household Relocations on Children's Academic Achievement," *Journal of Labor Economics*, Vol. 24, No. 2, April 2006, pp. 319–350.

Mattock, Michael G., and Jeremy Arkes, *The Dynamic Retention Model for Air Force Officers: New Estimates and Policy Simulations of the Aviator Continuation Pay Program*, Santa Monica, Calif.: RAND Corporation, TR-470-AF, 2007. As of August 29, 2012: http://www.rand.org/pubs/technical_reports/TR470.html

Public Law 106-65, National Defense Authorization Act for Fiscal Year 2000, October 5, 1999. As of August 29, 2012: http://www.gpo.gov/fdsys/pkg/PLAW-106publ65/html/PLAW-106publ65.htm

R Development Core Team, *R: A Language and Environment for Statistical Computing*, computer language, Vienna, Austria: R Foundation for Statistical Computing, 2005.

Savych, Bogdan, *Effects of Deployments on Spouses of Military Personnel*, Santa Monica, Calif.: RAND Corporation, RGSD-233, 2008. As of July 8, 2013: http://www.rand.org/pubs/rgs_dissertations/RGSD233.html

U.S. Air Force, *Fiscal Year 2009 Aviator Continuation Pay (ACP) Program*, 2009.

U.S. Department of Defense, *Report of the Tenth Quadrennial Review of Military Compensation*, Volume I: *Cash Compensation*, February 2008a. As of August 29, 2012: http://www.whs.mil/library/doc/Tenth.pdf

———, *Report of the Tenth Quadrennial Review of Military Compensation*, Volume II: *Deferred and Noncash Compensation*, July 2008b. As of August 29, 2012: http://www.defense.gov/news/QRMCreport.pdf